Grape Pest Management

Technical Editors

Donald L. Flaherty, Chairman
Entomology Farm Advisor, Cooperative Extension,
Tulare County Farm Advisors' Office

Frederik L. Jensen
Viticulturist, Cooperative Extension,
San Joaquin Valley Agricultural Research and Extension Center, Parlier

Amand N. Kasimatis
Viticulturist, Cooperative Extension, Davis

Hiroshi Kido
Staff Research Associate, Entomologist, Department of Entomology, Davis

William J. Moller*
Plant Pathologist, Cooperative Extension, Davis

Editorial Production

Heidi Seney
Editor, Agricultural Sciences Publications

Vonzetta Gant
Graphic Artist, Agricultural Sciences Publications

Jack Kelly Clark
Photographer, Cooperative Extension

Mel Gagnon
Communications Specialist, Cooperative Extension

---

*Deceased June 23, 1981.

Originally published in 1981. Edition of 1982 slightly revised.

Copyright © 1981 by The Regents of the University of California
Library of Congress Catalog Card Number: 80-70846
International Standard Book Number: 0-931876-44-3

Printed in the United States

4m-5/82-HS/VG

GRAPE PEST MANAGEMENT

Contents

NOTE: Detailed contents pages will be found before each of the nine sections listed above.

PRECAUTIONS FOR USING PESTICIDES

Pesticides are poisonous and must be used with caution. *READ THE LABEL BEFORE OPENING A PESTICIDE CONTAINER.* Follow all label precautions and directions, including requirements for protective equipment. Use a pesticide only against pests specified on the label or in published University of California recommendations. Apply pesticides only on the crops or in the situations listed on the label. Apply pesticides at the rates specified on the label or at lower rates if suggested in this publication. Laws, regulations and information concerning pesticides change frequently, so be sure the publication you are using is up to date.

Responsibility: The user is legally responsible for any damage due to misuse of pesticides. Responsibility extends to effects caused by drift, runoff or residues.

Transportation: Do not ship or carry pesticides together with food or feed in a way that allows contamination of the edible items. Never transport pesticides in a closed passenger vehicle or in a closed cab.

Storage: Keep pesticides in original containers until used. Store them in a locked cabinet, building or fenced area where they are not accessible to children, unauthorized persons, pets or livestock. DO NOT store pesticides with foods, feeds, fertilizers or other materials that may become contaminated by the pesticides.

Container disposal: Dispose of empty containers carefully. Never reuse them. Make sure empty containers are not accessible to children or animals. Never dispose of containers where they may contaminate water supplies or natural waterways. Consult your County Agricultural Commissioner for correct procedures for handling and disposal of large quantities of empty containers.

Protection of nonpest animals and plants: Many pesticides are toxic to useful or desirable animals, including honeybees, natural enemies, fish, domestic animals and birds. Crops and other plants may also be damaged by misapplied pesticides. Take precautions to protect nonpest species from direct exposure to pesticides and from contamination due to drift, runoff or residues. Certain rodenticides may pose a special hazard to animals that eat poisoned rodents.

Posting treated fields: For some materials re-entry intervals are established to protect field workers. Keep workers out of the field for the required time after application and, when required by regulations, post the treated areas with signs indicating the safe re-entry date.

Harvest intervals: Some materials or rates cannot be used in certain crops within a specified time before harvest. Follow pesticide label instructions and allow the required time between application and harvest.

Permit requirements: Many pesticides require a permit from the County Agricultural Commissioner before possession or use. When such materials are recommended in this publication, they are marked with an asterisk (*).

Processed crops: Some processors will not accept a crop treated with certain chemicals. If your crop is going to a processor, be sure to check with the processor before applying a pesticide.

Crop injury: Certain chemicals may cause injury to crops (phytotoxicity) under certain conditions. Always consult the label for limitations. Before applying any pesticide, take into account the stage of plant development, the soil type and condition, the temperature, moisture and wind. Injury may also result from the use of incompatible materials.

Personal safety: Follow label directions carefully. Avoid splashing, spilling, leaks, spray drift and contamination of clothing. *NEVER* eat, smoke, drink or chew while using pesticides. Post medical care information and provide for emergency medical care *IN ADVANCE* as required by regulation.

To simplify information, trade names of products have been used. No endorsement of named products is intended, nor is criticism implied of similar products which are not mentioned.

INTRODUCTION

Integrated pest management (IPM) in grapes had its beginnings in the late 1950s. By that time the grape leafhopper, *Erythroneura elegantula* Osborn, had developed considerable resistance to the new synthetic organic insecticides, and there were accompanying biological upsets of spider mites and grape mealybug. University of California Agriculture Experiment Station entomologists and Cooperative Extension viticulturists, with the active support of the grape industry, began intensive studies to lay the groundwork for integrating chemical, cultural and biological controls into a practical pest management program. A few growers and vineyard managers quickly adopted the new research findings and grape IPM was under way.

Studies, for example, showed that large acreages of grapes planted near streams and rivers, where wild grapes and wild blackberries (*Rubus* spp.) flourish, seldom require control for grape leafhopper. This is because of the activity of a minute wasp, *Anagrus epos* Girault, which parasitizes the eggs of blackberry leafhopper, *Dikrella californica* Gillette, a noneconomic species whose eggs are present throughout the year on wild blackberries. Survival of the parasite depends upon the presence of the *Dikrella* leafhopper eggs because grape leafhopper eggs are not present during winter.

Additional accomplishments in the late 1960s included the determination of treatment levels for various insect and spider mite pests; development of more efficient ways of applying pesticides; adoption of cultural practices that favor abiotic or biotic natural controls; and use of selective pesticides to reduce the problems of biological upsets.

However, in spite of publication and demonstration of most of these findings, widespread adoption of integrated control by the grape industry did not occur. Compared with the difficulty of acquiring knowledge and experience to implement long-range integrated pest management programs, it has been easier for too long to solve vineyard pest problems in the short run by relying on pesticides. Consequently, industrywide integrated pest management programs in grapes were held in abeyance through the late 1960s and most of the 1970s. But with the advent of inflated costs of pest control and greater concern for the environment came wider interest in IPM and the recognition that a clear understanding of procedures will help make pest management work more easily.

This grape IPM book is an outgrowth of that widespread interest and the increasing need to provide better control methods.

While past grape IPM information in California came primarily from studies in the San Joaquin Valley and dealt entirely with insect and spider mite problems, this publication looks at problems statewide and considers *all* pests that are destructive of grapes: diseases, nematodes, vertebrates and weeds, as well as insects and mites. But because pest management is ever changing with the development of new sampling and monitoring techniques, new pest outbreaks, new pest resistance problems, etc., we must emphasize that this book is to be used only as a guide for grape IPM. Growers and pest control advisers must still determine for themselves the procedures most applicable to their particular vineyard situations.

ACKNOWLEDGMENTS

Developing the information for a publication of this nature requires knowledge from many disciplines and the close cooperation of many people.

Growers contributed the use of their vineyards for trials and studies. In doing so, some incurred losses in order that integrated pest management (IPM) methodology could be developed. Moreover, many growers contributed their own astute observations of various pest problems.

Private consultants and pest control advisers contributed practical applications of IPM research findings.

Industry, USDA and U.C. research scientists spent many hours in laboratory and field, going back and forth to examine findings and to rework data until the information they had sifted through became useful to the grower.

The University of California's Cooperative Extension farm advisors and specialists spent many hours working with Experiment Station scientists to develop grape IPM and as many hours to teach its principles.

Many of these same people were active in assembling *Grape Pest Management*. It is never easy to single out all of those who contributed to a publication, but we would be remiss if we did not note for starters our debt to the forerunner of this book, *Pear Pest Management*. It was our model for organizing this book and we are particularly grateful to Richard S. Bethell, its technical editor, and all those who assisted him.

In connection with our own project, we would like to acknowledge the efforts of:

Jack Kelly Clark, Cooperative Extension photographer, whose color photographs of the flora and fauna in the world of grapes have enriched this book. Except where noted, all photographs in this book were taken by Jack.

Mel Gagnon, Cooperative Extension communications specialist, who helped us organize and write this book so that it will be easy for you to use.

Vonzetta Gant, Agricultural Sciences Publications artist, whose arrangements of text and photographs make this publication handsome and readable.

Heidi Seney, Agricultural Sciences Publications editor, who guided the assembling of this book with patience and care.

Grape Pest Management would not have been realized without the foresight and dedication of a number of grape research scientists who initiated early grape integrated pest control studies. Much of the information is taken from their published writings and personal observations. We would like to express our gratitude to: Dr. Luigi Chiarappa, formerly with DiGiorgio Fruit Corporation and now with the Food and Agriculture Organization, Rome, Italy; Professor Emeritus Richard L. Doutt, U.C., Berkeley; Professor Emeritus William B. Hewitt, U.C., Davis; Professor Carl Huffaker, Division of Biological Control, U.C., Berkeley; Dr. James R. Cate, Jr., Texas A&M University; Mr. John Nakata, staff research associate, now retired, Kearney Field Station, Parlier; the late Professor Emeritus Eugene Stafford, U.C., Davis; and Dr. Victoria Y. Yokoyama, California State University at Long Beach.

Finally, we wish to express our thanks to the University of California, Division of Agricultural Sciences, its Cooperative Extension and Experiment Station, and to the California Department of Food and Agriculture for support and contributions to the successful development of *Grape Pest Management*.

Donald L. Flaherty, Chairman, Technical Editors

HOW TO USE THIS BOOK

This publication provides detailed information about more than a dozen common diseases, more than 40 insects and mites, a dozen or more nematodes, about a dozen vertebrates, and many annual and perennial weeds. All are pests and their common targets are raisin, wine and table grapes. Their yearly damage and cost of control in California's vineyards run into millions of dollars. Moreover, because the grape is the number one fruit crop in California, bringing to it an annual revenue in the neighborhood of $1 billion, it would seem wise that vineyard managers learn about the least expensive and most efficient means to curb destruction by pests.

With so many possibilities for destruction of the grape crop, it behooves growers and pest control advisers to consider them carefully when making decisions about how to bring about the best production in vineyards. It also behooves the grower and adviser to recognize the importance of close coordination between vineyard cultural operations and pest control practices.

Among the most valuable guides in this book, the authors hope, are five calendars on viticulture management for the five major grape-growing areas of California: the north coast, central and south coast, northern San Joaquin Valley, southern San Joaquin Valley and Coachella Valley. These give the grower and pest control adviser the general timing for basic viticulture practices along with monitoring and treatment decision timing for the major pests in the area. The charts are to be found in *Section I. General Viticulture.*

The rest of the book is divided into sections on diseases, major insect and mite pests, minor insect and mite pests, stored raisin products pests, nematodes, vertebrate pests, weeds, and pesticides, their application and safety requirements. Each section is organized to help you find specific information. For example, the subtopics in the section on diseases are: symptoms, disease cycle and management of the disease. The sections on major and minor insects and mites begin with general information followed by pest description, injury, seasonal development, natural control, monitoring guidelines and management guidelines. Each section has its own table of contents.

With the exception of the weed section, where application rates are critical, no specific chemical recommendations (rates per acre, etc.) are given in this publication. This information is often subject to change and to include it would result in too many revisions to keep the book up to date. Therefore, in the management guideline sections, chemical recommendations are only given in a broad sense to let the reader know what kinds of materials best fit into integrated pest management programs for grapes. Pesticide labels should always be consulted for specific information.

Finally, the goal of this book is to provide a "state of the art" view of the pests that affect grape production. Much needs to be learned, as evidenced by the fact that a great deal of information is given for some pests and little for others. Pest management is a dynamic process. When you use this publication, regard it as a contribution to an evolutionary process rather than as a dictum. It is hoped that this contribution will prove beneficial to those whose aim is to increase the success of integrated pest management for grapes.

Pesticide registration and use requirements are subject to change. To insure proper chemical usage, contact your local farm advisor or County Agricultural Commissioner.

On the preceding page:

It takes effort to grow beautiful, marketable grapes for wine, table and raisins. Two viticultural practices are pictured: pruning (upper left photo) and sprinkler irrigation (lower left photo). One result of good viticulture practices is a beautiful cluster of Ruby Seedless (upper right photo). Lower right photo shows a typical table grape packing scene.

Section I—GENERAL VITICULTURE

Contents

ANNUAL GROWTH CYCLE
OF A GRAPEVINE

The annual cycle of growth of a grapevine is characterized by a grand period of shoot growth during late spring, flowering and the development of berries during late spring and early summer, continued fruit development during summer, rapid accumulation of sugars in the fruit during the ripening period, fruit maturity and harvest in fall, leaf senescence, and finally dormancy after frosts kill the leaves in late fall.

The emergence of growth in spring is temperature dependent and occurs after mean temperatures exceed 10°C (50°F). Development is often slow during this period because weather patterns are not stable. Once daytime temperatures warm up, growth proceeds at a more rapid pace reaching a climax about the time that the flowers begin to open during mid-May to early June.

It is during this period that a transition occurs in the flow of photosynthates. Whereas rapid shoot growth was the active "sink" for leaf-elaborated carbohydrates and other materials during spring, the tiny berries become a major sink with rapid enlargement after set, at a time when the rate of shoot growth declines. It is also during this period that a major flush of root growth occurs.

Berry development in grapes is similar to that of other fruits, characterized by three distinct stages: an initial period of rapid enlargement, followed by a short lag phase during which there is no gain in volume or weight, and climaxed by a rapid increase in size and marked increases in sugars, both leveling off during the stage of ripeness. Thus, the enlargement of berries can be characterized by a double s-shaped curve (see graph).

It is in the final stages of ripeness that the shoots become lignified and attain a brown color, often characteristic of the variety. This process of lignification begins near the base of the cane and continues distally with time.

A second minor flush of root growth also occurs in fall during harvest. After fruit removal the flow of photosynthates is largely directed to storage as carbohydrates in the arms, trunk and roots. These reserves are essential to vine survival during periods of extreme cold temperatures in winter. They are also essential to early development of shoots as vines leave dormancy in the spring.

Dormant bud.

Swollen buds photographed March 22, 1979.

Bud burst photographed March 22, 1979.

Early shoot growth photographed March 28, 1979.

GROWTH OF A GRAPE BERRY

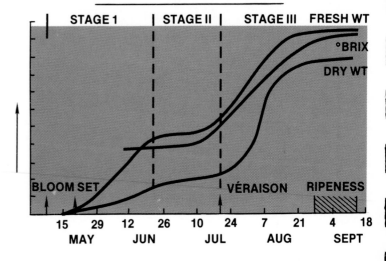

Flat-leaf stage photographed April 2, 1979.

Six-inch shoot growth stage photographed April 12, 1979.

"Sap balls" or "pearls," a natural exudate, may be found on the undersides of grape leaves in the spring; they are not to be confused with insect or mite eggs. Photographed April 12, 1979.

Twelve-inch growth stage photographed April 19, 1979.

13

Twelve-inch growth stage. Note early development of axillary buds which form lateral shoots.

Vine growth at beginning of blossoming period; shoots are as long as three feet. Photographed May 13, 1979.

Flower cluster shown just before blossoming. Photographed May 13, 1979. In inset calyptra is beginning to dehisce on flower in center of photo.

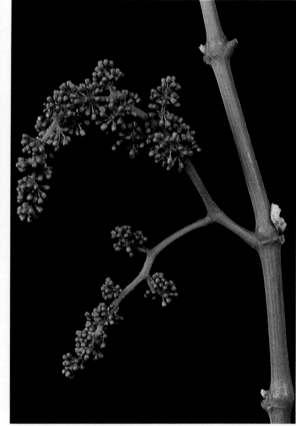

Flower cluster showing 1 percent calyptras off.

14

Flower cluster showing 35 percent calyptras off.

Flower cluster showing 10 percent calyptras off. In inset calyptra is in the last stage of dehiscing (left); after dehiscence, stamens begin to spread (right).

Flower cluster showing 85 percent calyptras off. In inset flower shows five stamens spread out and stigma on the tip of the ovary. Note dehiscing calyptra on the left.

Fruit cluster showing an early stage of berry development after set. Photographed June 5, 1979. Inset shows closeup of berry.

Fruit cluster photographed on June 19, 1979 when the berries were 8-10 mm in diameter.

Fruit cluster at the beginning of the ripening period when berries were beginning to turn soft; the soluble solids level of the whole cluster was 6.9° Brix. Photographed July 27, 1979.

Fruit cluster at harvest time when the Brix level of this cluster was 21.5° Brix. Photographed September 7, 1979.

Vine growth at the time of harvest.

Photo by Amand W. Kasimatis

Early stages of defoliation showing the loss of basal leaves. Note the brown, lignified canes indicating wood maturity. Photographed November 2, 1979.

In fall the leaves turn yellow and fall to the ground. Photographed November 2, 1979.

Dormant grapevine after leaf fall following killing frosts.

Dormant grapevine after pruning.

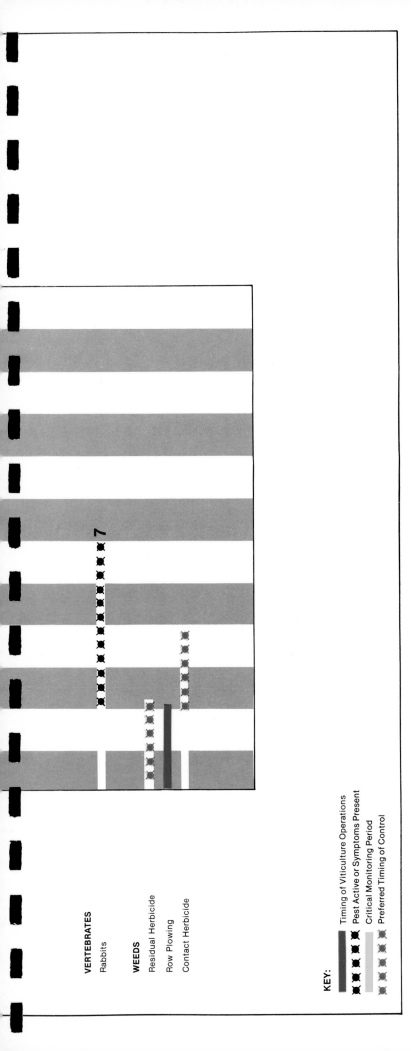

VERTEBRATES
Rabbits

WEEDS
Residual Herbicide
Row Plowing
Contact Herbicide

KEY:

Timing of Viticulture Operations

Pest Active or Symptoms Present

Critical Monitoring Period

Preferred Timing of Control

Color infrared aerial photography has been tested for several years as an aid in problem detection and management in selected vineyards in California. Its use can now be recommended as a routine tool—along with soil and plant tissue analysis, soil moisture measurement, and pest and disease control advising. Aerial photography should not be thought

of as a panacea, but rather as another key piece in the jigsaw puzzle of vineyard management.

Considerable experience has been gained in recognizing soil differences within a vineyard and the effect these have on irrigation management. Recognition of these soil differences should lead to a change

Photo by Jule Caylor, U.S. Forest Service

This aerial photo of about one square mile illustrates several types of vineyard problems that aerial photography can aid in detecting. At least three pests or disease problems and two soil problems are evident in the photo. The problems are discussed and illustrated in detail in these numbered photos: (a) Phylloxera, Photo 2; (b) oak root fungus, Photo 5; (c) fanleaf virus, Photo 8; (d) gravelly soils, Photo 9; (e) poorly drained soils, Photo 11.

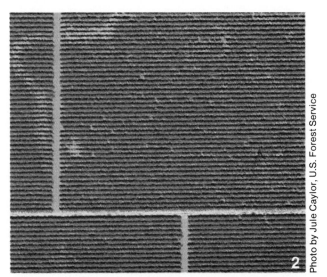

Photo by Jule Caylor, U.S. Forest Service

This spot infestation of phylloxera was first suspected when the aerial photo revealed a circular pattern of depressed growth. Root examination confirmed presence of the insect. Although phylloxera was known to exist on two margins of this vineyard, this spot was the first to be observed in the vineyard interior.

Photo by William E. Wildman

Vine in foreground is one of 12 that were depressed by phylloxera to create the spot on Photo 2. Unaffected vines are shown in the background.

in irrigation management which will result in improved yield and quality of grapes.

Pest and disease detection by color infrared aerial photography is usually only evident after the vines are irreversibly afflicted. However, this information is valuable in assessing relative thriftiness of the vineyard and determining what areas need special treatment or replanting.

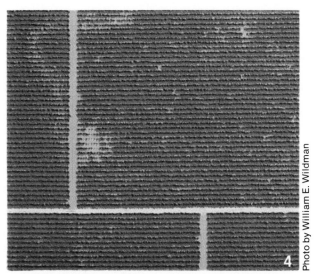

Photo by William E. Wildman

This aerial photo was taken one year later than Photo 2. The number of depressed vines had tripled to 37. Aerial photos will continue to be taken annually to search for new infestations and to make decisions on removal of infected blocks and their replacement by vines on phylloxera resistant rootstocks.

How to Obtain Color Infrared Aerial Photography

Good color infrared photography can probably give the most information about vine condition But ordinary color photography is often almost as good. In fact, any photography is more useful than none. But we will emphasize color infrared photography because we feel it has the most potential.

Color infrared film is available in 35 mm, 70 mm, 5 inch and 9 inch widths. 35 mm color infrared film can be used in any 35 mm camera that takes standard cartridges. A Wratten (or Tiffen) number 12 or 15 filter is also used to provide proper color balance. Therefore, if you want to take your own color infrared photos, 35 mm is the easiest way to go. 35 mm color infrared film is stocked at larger camera stores or can be specially ordered by smaller stores. It is designated Kodak Ektachrome infrared film IE 135–20 and is available in 20 exposure cartridges only. It is normally kept refrigerated before use and processed as soon as possible after exposure. Care should be taken not to allow the film to become hotter than normal room temperature during the use period.

Commercial operators often use 70 mm or 9 inch color infrared film. Its designation is Kodak Aerochrome Infrared film 2443. It is sold in reels of 100 feet or longer. These larger films produce photos with excellent definition, at a cost somewhat greater

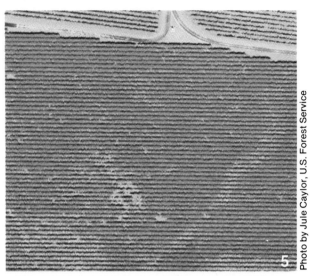

Photo by Jule Caylor, U.S. Forest Service

Oak root fungus has killed grapevines in an area where an oak tree was standing eight years before the aerial photo was taken. The vineyard was six years old when the photo was taken.

Photo by William E. Wildman

The dead area in Photo 5 looked much the same when photographed one year later, indicating that the spread of oak root fungus is relatively slow.

The pattern of Pierce's disease shown in this aerial photo is typical. Vines closest to the habitat of the sharpshooter leafhopper—the grassy and weedy stream bed—have died, while those further from the stream are less affected.

than 35 mm film. But the size of the film is not as important as the use to which that film is put.

How to Interpret Aerial Photography of Vineyards

Color infrared film does not react to heat or temperature differences in plants. It does record the reflectance in the green and red bands of visible light, and in the band just slightly beyond the red band, called the near infrared band. Normal green plants

Photo by Jule Caylor, U.S. Forest Service

Fanleaf virus disease is responsible for the diminished and spotty pattern of foliage growth on the right side of the road in this aerial photo. Their duller vine color, compared with the vines to the left, indicates more advanced maturity. Both blocks are Cabernet Sauvignon, but contrary to appearances, vines on the right are three years older than vines on the left. The fanleaf virus host, *Xiphinema index* nematode, does not occur in the soil on the left, since this vineyard, planted six years before, was the first ever on this soil.

reflect green light but absorb most of the red light. Plants also reflect strongly in the near infrared band. The internal cell structure of the leaf is responsible for this reflectance of near infrared radiation, so conditions which alter this cell structure, such as water stress or disease, may lower the near infrared reflectance.

Interpretation of color infrared aerial photos depends on detecting differences in (1) size, shape and amount of foliage growth of vines, (2) color tone of vines or (3) color of soil.

The patterns in which these differences occur may be regular or random. If they are regular, they often are related to present cultural practices, establishment methods in the vineyard or even to management of a former crop. Examples of these patterns are: training or trellising of vines; present or former tillage practices; variety or rootstock differences; irrigation differences between blocks; nonuniform distribution of irrigation water within blocks; soil compaction loosened at regular intervals by trenching or ripping; old roadways or disturbed sites that have been planted over; carryover toxic effects from herbicides; fertilization irregularities; and many more.

If the patterns are random, they are more likely to be related to soil differences or to disease or insect problems. Whether regular or random, the causes for the differences must be diagnosed by a combination of aerial photo interpretation and ground checking, using all the information that is available on soil and water analyses, irrigation method and management, entomological surveys and plant disease assays, crop history, etc.

Once a condition has been diagnosed, its increase or decrease can be monitored by color infrared photography, nearby outbreaks of the same condition can be anticipated, and eradication or management practices can be devised to improve the growth of the vineyard. In the following sections color infrared aerial photos and color ground photos will be used to illustrate several pest, disease and soil conditions and how these differ from each other in their effect on vine foliage growth, vine color and soil color.

Differences in Vine Size, Shape, Foliage Growth

It is usually best to wait until at least midsummer to photograph vineyards, if differences in vine growth are the primary means of interpretation. Early in the season the reflectance of the young growth is usually obscured by the large amount of bare soil (or in some cases, cover crop) between the vine rows. By the time normal vines have developed a full canopy, those vines not making normal growth will appear as "thin" spots on the photo.

Four pest and disease problems will be discussed. The pattern that each makes is somewhat different from the others. The problems are grape phylloxera, oak root fungus (*Armillaria mellea*), Pierce's disease and the fanleaf virus/*Xiphinema index* nematode complex.

The outbreak of grape phylloxera pictured here (Photo 2) is typical of a condition that starts at a point source and spreads outward. Apparently a colony of phylloxera insects was introduced at that point by farm equipment or other means. At the time the photo was taken, 12 vines were suffering severely depressed growth allowing a spot of bare soil beneath them to be seen on the photo. Photo 3 shows a close-up of the still alive but depressed vines. One year later, 37 vines were similarly affected, as shown in Photo 4. Two years later several satellite outbreaks around the original one became evident. Apparently the migration of the insects from the original point is not uniform in all directions, and after populations have built up to a certain level, new areas of depressed growth appear in accelerating fashion.

Oak root fungus often creates a pattern of spreading outward in concentric rings, but in this case (Photo 5) the pattern is more irregular because the fungus is just beginning to kill grapevines in areas where several oak trees were standing a few years before. Probably the irregular distribution of oak roots is most responsible for the pattern of dying vines. If allowed to continue unchecked, the fungus would be expected to spread outward from all points of infection, eventually creating circular areas of dead vines. The spread of oak root fungus is relatively slow (Photo 6), compared with the rapid spread of phylloxera in one year's time noted in the first example.

Pierce's disease spreads in a different manner, and therefore creates a rather characteristic pattern. The disease itself is caused by a bacterium that occurs in several weed hosts, especially grasses and sedges occurring along the margin of streams. Sharpshooter leafhoppers, which are vectors of the bacteria, live in the weedy areas, acquiring the bacteria and spreading them as they migrate for a short way into the vineyard to feed. In doing so, they infect the vines closest to their habitat and those further out in the vineyard are less affected. This leaves the characteristic pattern shown in Photo 7, in which most vines close to the stream bed have died, with a gradation to more living and healthy vines in the vineyard interior.

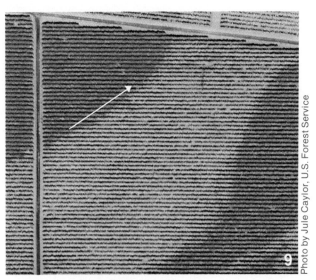

The sharp boundary in color of vines is caused by a gravel lens, which reduces the available water to the vines to the lower right of the arrow. This effect may not be noticeable in midsummer, but becomes very obvious in late September or October.

This low altitude aerial photo (from the roof of a car) was taken looking in the direction of the arrow in Photo 9. The boundary between the green and yellow vines is much less striking than in the aerial photo.

Fanleaf virus disease (Photo 8) shows up as an irregular, spotty pattern. The virus is carried by a nematode, *Xiphinema index,* which was introduced along with the virus into some of the best grape growing areas of California before the role of the nematode as a virus vector was understood. The disease does not kill vines, but it does cause them to be unfruitful and their growth to be irregular.

Differences in Color Tone of Vines or Soil

One needs to be especially careful when making interpretations based on the color tones on color infrared photos. In the first place, there are marked differences in green color (which translates to red color on color infrared film) among grape varieties. In the second place, the color tone of any given variety, even when healthy, will vary with the stage of growth, sun angle, extent of haze or cloudiness in the atmosphere, as well as with the photographic variability due to batch of film, processing and camera exposure. For practical purposes then, it is not possible to develop a series of exact color standards that will coincide with different growth conditions. And it is risky to make interpretations based on differences in color tones between different photos, particularly when taken on different dates.

Nevertheless, differences in color tone on the same photo can be useful in detecting relative differences in vine growth.

In general, a bright red or magenta color is characteristic of healthy, turgid vines that are making rapid shoot growth. Vines that are not making new shoot growth or are water-stressed appear a duller or darker red color. These color differences may be easily seen or may be subtle, and for a complete interpretation, it is necessary to check the vine status on the ground. Keep in mind that the bright color may be desirable early to mid-season when shoot growth should be vigorous, but in the latter part of the growing cycle, the duller colors are more desirable, because this indicates that the vine is putting its photosynthate into the fruit and wood, rather than into excessive shoot growth.

Late in the season, usually September or October, very striking differences in vine color can occur because of differences in soil waterholding capacity. Where vines have run short of water because of shallow root zones or coarse soil textures, leaves turn yellow. Color infrared photos show these leaves as a light straw color which contrasts strongly with the red color (on infrared film) of leaves that are still green. Often the boundaries are very sharp as shown by the arrow on Photo 9. This is largely a vine color difference and not a soil color difference. A layer of gravel in the subsoil limits the root depth and lowers the amount of water available to the vines. The ground photo in Photo 10 was taken looking in the direction of the arrow on Photo 9. The aerial photo gives a much clearer idea of the magnitude and extent of the soil difference.

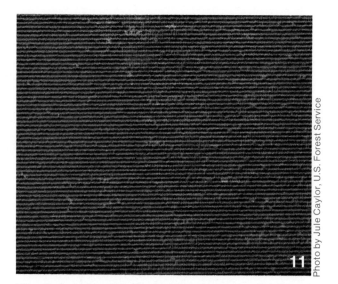

Photo by Jule Caylor, U.S. Forest Service

The dark colored soil showing between defoliated vines suggests a poorly drained clay soil.

Photo by William E. Wildman

A ground view of the poorly drained area in Photo 11 shows early defoliation from potassium deficiency that was induced by the high water table. Tile drainage has since been installed in this area.

Soil color also varies across a vineyard and can be indicative of differences in soil moisture, soil texture, parent material, drainage or depletion of topsoil. Light soil color often means sandy soils with good drainage. Dark gray or black soils contain more organic matter, which is usually related to clay texture or poor drainage or both. Eroded soils often are detected by the more reddish or brownish subsoil showing up where topsoil has been removed.

Photo 11 is an example of a poorly drained clay soil. Photo 12 shows this same area on the ground. Because of the high water table in this area, vines have defoliated, allowing more of the dark colored soil to show through.

Conclusion

Color infrared aerial photography is very useful in diagnosing problem conditions in vineyards and in devising management solutions to problems. The examples given here are only a few of the many conditions that can be recognized with relative ease by studying aerial photos. Many more possibilities for using aerial photos for problem diagnosis exist in vineyards and in other crops. Successful interpretation of the photos and development of management practices requires that the interpreter—whether a pilot-photographer, agricultural consultant or vineyard manager be well trained and experienced in viticulture, soils, irrigation, climates, entomology, plant pathology, etc. Only by combining all of his/her knowledge and experience with the results of other diagnostic tests will the vineyard manager be able to get the most out of aerial photography

ACKNOWLEDGMENT

The authors greatly appreciate the use of several aerial photos taken by U.S. Forest Service Photographer Jule Caylor, in conjunction with the *Seventh Biennial Workshop on Color Aerial Photography in the Plant Sciences*, held in Davis, California, May 15–17, 1979.

Viticulture

Bark—Tough covering of a woody stem or root external to the cambium.

Berm—Ridge of soil in the vine row.

Blade—Expanded portion of a leaf.

Bloom—(1) Flowering as indicated by shedding of the calyptras. (2) The waxy coating on grape berries; gives a frosted appearance to dark-colored varieties.

Bud—Rounded organ at the node of a cane or shoot containing an undeveloped shoot protected by overlapping scales. There are typically three buds at each node, a more developed primary bud between less prominent secondary and tertiary buds. *Figure I.*

    Basal buds—Small buds that lie at the base of a cane or spur as part of a whorl.

    Count buds—Number of buds on a spur or cane not including basal buds. First count bud has 1/4 inch or more separation from a basal bud below. Transitional forms may make determination of first count buds difficult.

Bud break—Stage of bud development when green tissue becomes visible.

Callus—Parenchyma tissue that grows over a wound or graft and protects it against drying or other injury.

Calyptra—Fused petals of the grape that fall off the flower at bloom.

Cambium—Thin layer of undifferentiated meristematic tissue between bark and wood. When active, it divides to give rise to the secondary tissues, xylem and phloem, resulting in growth in diameter of stems and roots.

Cane—Mature (woody) shoots. See **Shoot.**

Cap stem (pedicel)—Stem of individual flowers or berries.

Certified stock—Grapevine propagation material certified free of known virus diseases by the California Department of Food and Agriculture, under regulations of the Grapevine Certification and Registration program.

Chlorophyll—Green pigment of plants that absorbs light energy and makes it effective in photosynthesis.

Chlorosis—Yellowing or blanching of green portions of a plant, particularly the leaves, which can result from nutrient deficiencies, disease, herbicide injury or other factors.

Clone—Group of vines of a uniform type propagated vegetatively from an original mother vine.

Cropping

    Crop—Amount of fruit borne on vines.

    Crop load—Amount of crop in relation to vine's leaf surface.

    Crop recovery—Crop produced from new growth following injury by spring frost.

Crossarm—Horizontal or slanting crosspiece, usually 18 to 42 inches long, attached at or near the top of the stake to extend the width of the trellis. Wires are attached at each end of the crossarm. Additional wires may be added between the outer wires on the wider crossarms.

Crown—Point at or just below soil surface where main stem (trunk) and root join. Also sometimes used by growers synonymously with the term, head, but this use should be discouraged to conform with common horticultural terminology.

Crown suckering—See **Shoot thinning.**

Cultivar—A cultivated variety.

Cutting—A portion of dormant cane, usually 14 to 16 inches long, used for propagation; may also refer to a shoot section to be propagated under mist.

Degrees Brix—A measure of the total soluble solid content of grapes, approximately the percentage of grape sugars in the juice.

Deshooting—See **Shoot removal.**

Disbudding—Removal of buds or very young shoots (less than 6 inches long).

Dormant—(1) Plants, buds or seeds not actively growing. (2) Period between normal leaf fall and resumption of growth in spring.

Drop—Abscission of flowers after bloom. When completed, fruit set stage has been reached.

Epinasty—Downward bending of leaves caused by some hormone sprays, water stress, etc.

Eye—Compound bud of a grape.

Fertilization—(1) Application of mineral plant nutrients to soil. (2) During fruit set, the union of sperm cells from the pollen tube with egg cells of the ovary.

Field capacity—Amount of water retained in a soil against the force of gravity, usually measured 24 to 36 hours after irrigation, and about equal to the moisture content at one-third the atmospheric pressure. Also referred to as water-holding capacity.

Filament—Stalk supporting the pollen-bearing stamen in a flower.

French plowing—See **Row plowing.**

Fruit—Mature ovary (berry) or cluster of mature ovaries.

Fruit set—Stage of berry development one to three weeks after blossoming when most flowers have fallen and those remaining have set and develop into berries. Also called berry set stage or shatter stage; fruit set is preferred.

Gibberellin(s)—Plant growth regulator currently used for bloom thinning in Thompson Seedless, berry enlargement of seedless varieties and reduction of berry shrivel in Emperor.

Girdling—Removal of a complete ring of outer and inner bark from a shoot, cane or trunk. Also called ringing. This process temporarily interrupts the downward translocation of metabolites.

Glabrous—Hairless plant surface.

Grass culture—A type of management in which volunteer vegetation, mostly grasses, is allowed to grow without cultivation from late spring until sometime before or beyond harvest. Weeds are mowed periodically to control height. Grass culture differs from sod culture in that the soil is cultivated at least once a year and usually two or three times.

Green-manure crop—A crop grown and plowed under while still green to improve the soil, especially by adding organic matter.

Head—Upper portion of a vine consisting of the top of the trunk and arms. Sometimes the head is called the crown by growers; this usage is discouraged, because in common horticultural terminology the crown is at the ground surface.

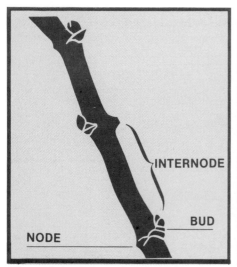

Figure 1. Location of the main features of the cane.

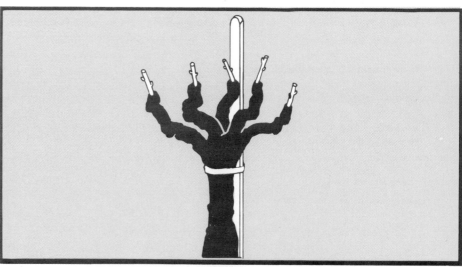

Figure 3. A head-trained vine with spur pruning.

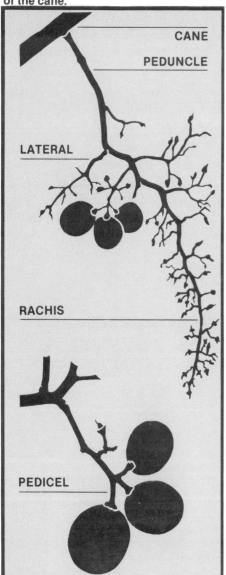

Figure 2. Structure of the grape cluster and its attachment to the cane.

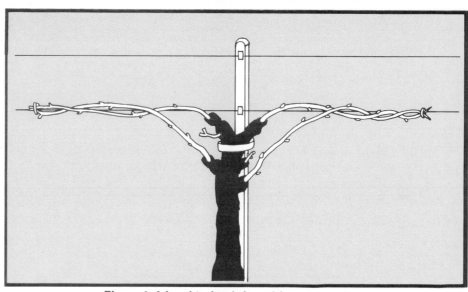

Figure 4. A head-trained vine with cane pruning.

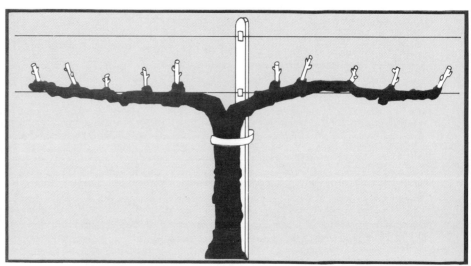

Figure 5. Bilateral cordon training with spur pruning.

38

Indexing—Determination of the presence of disease in a vine by removing buds or other parts and grafting them onto a readily susceptible plant that exhibits symptoms of a transmissible disease.

Inflorescence—Flowering cluster of the grapevine.

Internode—Section of a shoot or cane between two adjacent nodes. *Figure I.*

Latent bud—Bud that has remained undeveloped for a season or longer.

Lateral—Branch of the main axis of a cluster (*Figure 2*). Also a side shoot arising from the main shoot.

Layer—Long cane from an adjacent vine utilized to replace a missing vine.

Leaf—Relatively flat, thin expanded organ growing from the shoot consisting of a broad blade, a petiole and two inconspicuous stipules at the base of the petiole.

Leaf scar—Scar left on a cane after leaf fall.

Lenticel—Tiny porelike opening surrounded by corky tissue, often prominent on grape berries and pedicels.

Margin—Edge of a leaf blade.

Meristem—Undifferentiated tissue, the cells of which are capable of active cell division and differentiation into specialized tissues.

Node—Enlarged portion of a cane or shoot at which leaves, clusters, tendrils and/or buds are located at regular intervals. *Figure I.*

Ovary—Enlarged basal portion of the pistil containing ovules or seeds.

Own-rooted—Vine grown from a cutting that develops its own root system as opposed to a vine grafted or budded onto a rootstock.

Pedicel—Stalk of an individual flower or berry in a cluster. *Figure 2.*

Peduncle—That portion of the rachis (cluster stem) from the point of attachment to the shoot to the first lateral branch on the cluster. *Figure 2.*

Petiole—Leaf stalk attaching the leaf blade to the shoot.

pH—Refers to degree of acidity or alkalinity as a scale of numbers from 1 (very acid) to 14 (very alkaline). pH 7.0 is neutral, representing the reciprocal of the hydrogen ion concentration and expressed in gram atoms per liter of a solution.

Phloem—Region of tissue in the plant composed of sieve tubes and parenchyma which translocates food materials elaborated by the leaves.

Photosynthesis—Process by which a plant converts carbon dioxide and water into carbohydrates. Solar radiation is the energy source for this process.

Phytotoxic—Causing injury or death of plants or portions of plants.

Pistil—Female part of the flower, consisting of a stigma, a style, and an ovary.

Pollination—Transfer of pollen from anther to stigma.

PPM—Parts per million. Concentration of a material expressed as the number of units per million units. It is the same as milligrams per liter.

Pruning—*Figures 3, 4, 5.*

    Cane—Pruning method whereby canes are retained as fruiting units.

    Spur—Pruning method whereby spurs are retained as fruiting units.

Rachis—Main cluster stem including peduncle. *Figure 2.*

Ringing—See **Girdling.**

Rooting—A young vine produced from a cutting grown for one season developing both roots and shoots.

Rootstock—Specialized stock material to which fruiting varieties of grapes are grafted to produce a commercially acceptable vine. Grape rootstock varieties are used for their tolerance or resistance to root parasites, such as phylloxera and nematodes, or for vigor.

Row plowing—Specialized type of plowing that removes the narrow ridge of soil (berm) and weeds in the vine row. Plow avoids vines and stakes in a row by means of a mechanical or hydraulic lever. A pass on each side of the row is required.

Scion—Fruiting variety that is grafted or budded onto a stock.

Serration—Toothlike indentation at a leaf margin.

Shatter—Detachment of berries from cluster, either with or without the pedicel. Has been used to indicate fruit set but should be reserved for detachment of berries after this stage.

Shelling—Abscission of flowers before or in bloom. Occurs previous to drop after bloom.

Shoot—Current season's succulent and green stem growth. It becomes a cane when more than half of its length becomes woody as indicated by tan or brown color.

Shoot removal or deshooting—Removal of unwanted shoots on the trunk of a vine below the head.

Shoot thinning—Removal of unwanted shoot growth from the head, cordon or arms of a vine when the shoots are short, usually 6 to 15 inches long. (Six-inch or shorter shoots covered under **Disbudding**.)

Shot berries—Very small berries that fail to develop to normal size; usually seedless.

Sinus—Cleft or indentation between the lobes of a leaf blade.

Sod culture—Type of management in which a permanent ground cover is kept at all times and is usually mowed periodically during the growing season.

Spur—A short fruiting unit of one-year growth, usually consisting of one or two nodes; retained at pruning.

Stamen—Pollen-producing organ of a flower, consisting of the anther and a filament.

Stigma—Upper surface of pistil, where pollen grain is received and germinates.

Stock—See **Rootstock.**

Stomate (stomata)—Tiny opening(s) bordered by two guard cells in the epidermis of a leaf or young stem which regulates the inward flow of carbon dioxide and the outward flow of water vapor lost in transpiration.

Stylar scar—Small corky area remaining at the apex of a berry after the style dries and falls off following blossoming.

Style—Portion of pistil between stigma and ovary.

Sucker—Shoot arising at or below ground level.

Suckering—Removal of shoots arising at or below ground, but a term often used by growers to include removal of unwanted shoots arising on trunk.

Tendril—Slender twining organ on a shoot opposite a leaf that can coil around an object for support.

Tomentum—Growth of short, matted, wooly hairs on leaves or stems.

Translocation—Movement of water, nutrients, chemicals or elaborated food materials within a plant.

Transpiration—Water loss by evaporation from leaf surface and through stomata.

Trellis—Permanent vine-supporting system consisting of stakes, wire and, often, crossarms.

Trunk—Main stem or body of a vine between roots and head of vine.

Variety—Group of closely related plants of common origin and similar characteristics within a species.

Véraison—Beginning of fruit ripening.

Vine training

> **Bilateral cordon**—*Figure 5*. System of vine training that divides the trunk into two permanent branches, each extending in opposite directions down the vine row and horizontally supported on a trellis wire; commonly referred to as cordon training. Vines are spur-pruned.

> **Head**—*Figures 3, 4*. Simple system of vine training in which upright trunk is held by a stake; it terminates in short permanent arms that bear spurs or canes.

Wing—Well developed basal cluster branch, appearing separated from main cluster.

Xylem—Woody portion of conducting tissue whose function is to conduct water and minerals.

On the preceding page:

Powdery mildew is the number one disease of grapes in California. Pictured is an early season infection on a young shoot.

Section II—DISEASES

Contents

POWDERY MILDEW

Scarring on canes resulting from shoot infection.

Mildew colonies on leaf's underside are first seen as small brown patches.

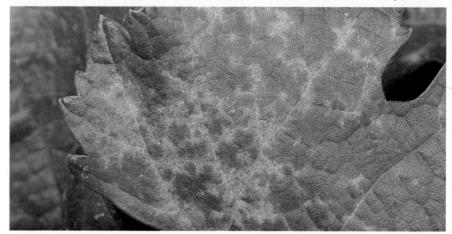

Mildew colonies on upper leaf surface appear as small white fuzzy patches.

Heavy mildew infection on Thompson Seedless showing berry stunting.

Heavy mildew infection on Cabernet Sauvignon showing delay in coloration.

Powdery mildew infection on a mature Cabernet Sauvignon cluster showing formation of cleistothecia.

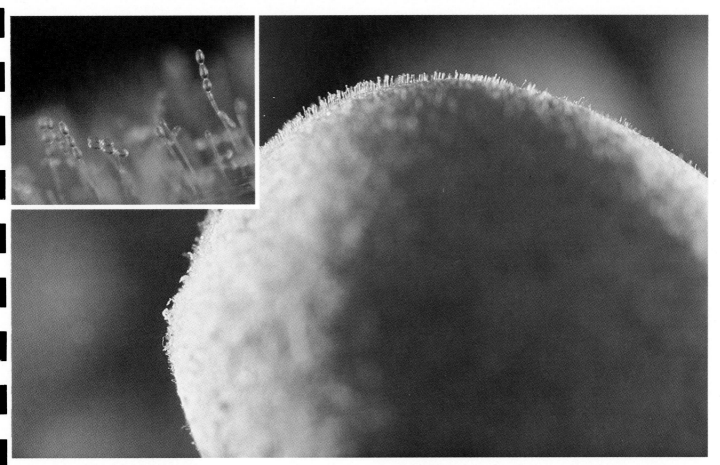

Mildew spores are formed on short stalks. Inset shows spores borne in chains of three or four.

Weblike scars on Thompson Seedless.

Cap stem infections.

Development of cleistothecia in late summer. Inset shows mature cleistothecium.

Powdery mildew, caused by the fungus *Uncinula necator*, is perhaps the most enduring and widespread problem in California *Vitis vinifera* grapevines. The amount spent on mildew control, plus the loss in yield, can frequently equal 10 percent of the state's entire grape crop value. Despite regular control measures, the disease can result in even heavier losses in highly susceptible varieties in some years. Early season mildew interferes with fruit set and development. Later, severe mildew may cause berry cracking, allowing rot organisms to enter.

The degree of susceptibility to mildew varies. Carignane, Thompson Seedless, Cardinal, Chardonnay, Cabernet Sauvignon and Chenin blanc are seriously affected, while Petite Sirah, Zinfandel, Semillon and White Riesling are less susceptible.

The fungus also infects other members of the plant family *Vitaceae*, including all species of native American grapes: *Vitis labrusca, V. aestivalis, V. rotundifolia, V. rupestris, V. vulpina, V. californica* and others. These species are less severely affected than the *V. vinifera* grape commonly grown in California. Other related susceptible species are: monk's hood vine (*Ampelopsis aconitifolia*) and Virginia creeper or Boston ivy (*Parthenocissus quinquefolia*).

Symptoms

All succulent tissues on a grapevine are susceptible to mildew infection and show characteristic symptoms. In some vineyards (especially Carignane), young shoots entirely covered with mildew can be found shortly after bud break. More commonly, the disease is first observed on young berries growing under a heavy vine canopy. Infected berries become scarred and later may crack open, allowing entry of rot organisms. The fungus forms a white, webby mat of strands (mycelium) over the infected tissue's surface. Short rootlike branches (haustoria) grow from the mycelium into the uppermost layer of plant tissue to draw out nutrients. Chains of spores (conidia) borne on short stalks arise from the mycelium, giving a dusty or powdery appearance to the surface. Mildew colonies on leaves are usually found either on the underside of exposed leaves or on both sides of well shaded leaves. These colonies can be detected at any early stage in their development by faint yellow patches about 6 mm (¼ inch) in diameter and observ-

ing the associated characteristic webbing and spore chains with a hand lens (10X).

Late in the season (usually after mid-August), small, spherical, black fruiting bodies (cleistothecia) may be formed amidst the mycelial mats; they contain a second type of spore (ascospore). If mildew was extensive in the previous season, look for red speckled, stainlike scars of old infections on canes. The degree of staining observed in winter is a good indication of the level of shoot infection previously.

Disease Cycle

Powdery mildew cannot survive on dead grape tissue, so it survives the winter as dormant mycelium under infected buds or as ascospores in the cleistothecia on debris. Although cleistothecia are found frequently in California, there is no convincing evidence that the ascospores they contain will germinate and produce infections. Their role in the powdery mildew disease cycle is not known.

In the spring mycelium in infected buds grows over the emerging shoots, producing many spores. Therefore, the time when primary infections occur in a given location may depend upon the presence of, or proximity to, infected buds carrying over from last season.

Powdery mildew is further spread by windborne spores. Nothing is known about how far these spores can travel and remain infective or how many are produced per day. However, observations show infections spread most rapidly in a downwind direction if conditions are favorable. Thus, vineyards downwind from a severely infected vineyard need extra protection. The spread of powdery mildew is assisted by the presence of extensive grape plantings, backyard grapevines, wild grapes and other hosts, such as the ornamental Virginia creeper.

The spores land, producing new infections, and the cycle repeats itself many times during the growing season. The damage sustained by a vineyard depends largely upon the time of first infection. Early fruit infections cause stunted berries, scarring and off-flavors. Powdery mildew also may affect the rate of photosynthesis by infected leaves and, when vines are severely infected, it may impair their ability to produce adequate amounts of sugar.

The susceptibility of various plant parts to powdery mildew infection changes through the season. The fruit is susceptible to infection from the beginning of development until the sugar content reaches about 8 percent, which is why early season control is so important. Established infections then continue to produce spores until the berries contain 12 to 15 percent sugar. Old infections become inactive and the berries become immune after the sugar content exceeds 15 percent. Likewise, mildew on leaves develops best on young tissue and usually will not infect leaves over 2 months old unless they have been growing under dense shade. Shoots, petioles, and cluster parts are susceptible throughout the growing season. Because mildew requires a living host, dead or dormant tissues are not infected.

Powdery mildew is favored by mild weather. Spores germinate at leaf surface temperatures between 6° to 33°C (43° to 90°F), the optimum being 25°C (77°F). Rapid germination and mycelial growth take place at 21° to 30°C (70° to 86°F). At optimum temperatures, the generation time, that is the time between spore germination and production of spores by the new colony, is only five days. Leaf temperatures above 33°C (90°F) kill spores and mildew colonies. Thus, it is evident that grape powdery mildew is capable of developing at temperatures below those favorable for grape plant growth, but it is destroyed at high temperatures that do not harm the host plant.

The temperatures given in the previous discussion were measured at the leaf or fruit surface and may vary from 5° to 10°C (10° to 20°F) above or below the air temperature, depending on the intensity of the radiation to which the leaves or fruit are exposed. The lethal effect of air temperatures above 33°C (90°F) is buffered by the host plant. For instance, during short periods when air temperatures surpass 40°C (104°F), colonies exposed to sunlight and the full effect of high temperature would be killed rapidly, but colonies on cluster stems protected by cool berries under foliage might survive. The fungus is destroyed completely when air temperatures exceeding 35°C (95°F) last for extended periods in the early part of the growing season when the berries are small and the foliage is sparse. No accurate generalization can be made concerning the effect of air temperature on powdery mildew because so many factors relating to vine vigor and canopy density can

modify the temperature experienced by the mildew on the leaf surface.

Temperature plays a larger role in disease development than does moisture. Normal development of the disease can occur over the full range of relative humidity; however, free water, such as rain, dew or irrigation water, can cause poor and abnormal germination of spores and may wash spores and mycelium from the host tissue. One should not assume that water in the absence of a wetting agent will control powdery mildew. The spores and mycelium are somewhat hydrophobic and therefore are not easily "wet" by water; under dense canopies many may escape the influence of rain or wash water. Water lowers the temperature under the canopy and may actually enhance the rate of development of surviving infections, as is evidenced by occurrence of severe infections on Chardonnay and other varieties under sprinkler irrigation.

Management of the Disease

In general, the best control of powdery mildew is prevention. Sulfur continues to be the most effective and economical material, and most years regular applications of sulfur dust will provide adequate mildew control. But to be effective sulfur must be present before the fungus develops. Dusting should begin 14 days after bud break and should be repeated every 14 days. Or, if growth is rapid, applications should start when the shoots are 150 mm (6 inches) long, be repeated when shoots are 300 to 450 mm (12 to 18 inches) long, and then continued about every 14 days until fruit begins to ripen (wine and raisin grape vineyards) or until summer temperatures are too high for infection. If dusting every other middle, then the time interval should be shortened to seven to ten days. Wine grape and raisin grape growers are able to stop sulfuring when ripening begins because the berries cease to be susceptible after they achieve an 8 percent sugar content. However, to make sure all berries have reached this point, sulfuring should not be stopped until the average sugar test is 12 to 13 percent.

Table grape growers have to continue dusting as long as temperatures remain favorable for infection. This is to prevent infection of cap stems, which remain susceptible; infected cap stems result in poor storability. In the San Joaquin Valley sulfuring in table grapes is often discontinued during hot weather after early July, but it is resumed in late August when temperatures cool.

Because sulfur easily washes off vines, it should be reapplied after rain or sprinkler irrigation. Furthermore, optimum temperatures for mildew growth often follow rainstorms.

Sulfur prevents infection by mildew spores. It is not known whether the spores must be in direct contact with the sulfur particles to achieve control or whether a vapor phase of sulfur is toxic. In either case, good coverage is the key to effective control. There is no experimental evidence that sulfur can kill existing mildew colonies.

Sulfur can cause severe burns on vines if applied when air temperatures are near 38°C (100°F), especially in spring and early summer, so care should be exercised during hot spells. Burning may be lessened by (1) reducing the amount of sulfur used or (2) applying dust in early evening to allow slow oxidation during the night.

In years when conditions are particularly favorable for mildew, sulfur dust may not give adequate control. If a dusting is missed and infections become established, especially early in the season, additional dust applications may not control further spread of the disease. Under these conditions, use wettable sulfur plus a suitable wetting agent and water to wash off infections. The eradication effect is probably from the water and wetting agent; the wettable sulfur merely replaces the sulfur washed off during the application. The effectiveness of mildew eradication depends on the penetration and coverage of the vine canopy and clusters by the wash water. It is generally more effective when shoots are short and decreases as the canopy becomes denser. In severe cases several washings may be needed for adequate coverage and control once the canopy uniformly touches the vineyard floor.

The probability of severe mildew problems varies greatly between grape varieties and local climate types. Depending on the area, differing numbers of dustings are required to achieve an adequate control level, e.g., growers in the Napa Valley may use only half or fewer treatments than those required in the San Joaquin Valley. Local conditions must be taken into account in determining a mildew control scheme.

POWDERY MILDEW CYCLE

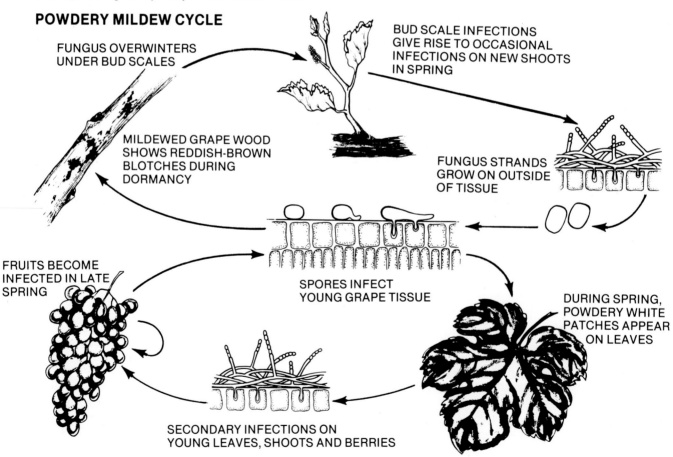

FUNGUS OVERWINTERS UNDER BUD SCALES

BUD SCALE INFECTIONS GIVE RISE TO OCCASIONAL INFECTIONS ON NEW SHOOTS IN SPRING

MILDEWED GRAPE WOOD SHOWS REDDISH-BROWN BLOTCHES DURING DORMANCY

FUNGUS STRANDS GROW ON OUTSIDE OF TISSUE

FRUITS BECOME INFECTED IN LATE SPRING

SPORES INFECT YOUNG GRAPE TISSUE

DURING SPRING, POWDERY WHITE PATCHES APPEAR ON LEAVES

SECONDARY INFECTIONS ON YOUNG LEAVES, SHOOTS AND BERRIES

Photo by Jeff Hall

Botrytis lesion at node of young shoot in spring.

Photo by William J. Moller

Botrytis blighted flower cluster.

Early spring Botrytis infection on leaf.

Rotting of young tissue characteristic of Botrytis shoot blight.

Single infected berry. Inset shows sporulation on cap stem and infected berry.

Lesion at base of young shoot.

Clusters of *Botrytis* spores on long stalks.

Destruction of central part of cluster caused by Botrytis.

"Raisining" caused by late season Botrytis infection.

Late season Botrytis infection. Desiccation and sporulation resulting from late season infections are seen in inset.

Sporulation on late season infection.

Overwintering Botrytis mummies.

Strands of fungal tissue develop on berry surfaces.

Spore production on mummies in early spring.

Bunch rots affect the yield and quality of wine, table and raisin grapes throughout California. In most years, statewide losses range from 1 to 10 percent; in certain varieties in a bad year, such as 1976, losses can reach 40 percent. While such devastating economic damage occurs only once every few years, the increasing value of grapes, as well as the introduction of chemicals that provide effective control, has increased grower interest in preventive measures.

Bunch rot can be caused by one or more of many different organisms. More than 70 species of fungi have been associated with fruit rots of grapes; most of them are secondary invaders capable only of infecting tissue that has been damaged in some way. A few organisms, however, are considered primary pathogens. In addition to attacking damaged fruit, they are capable of establishing infections in undamaged fruit when sufficient moisture is present.

BOTRYTIS BUNCH ROT
(Gray Mold)

The most important primary pathogen causing bunch rot is the fungus *Botrytis cinerea*. Much confusion is associated with this organism because in addition to causing heavy crop loss, it is responsible for the phenomenon of "noble rot"* under certain conditions. However, the proper sequence of climatic events leading to the occurrence of noble rot is not common, and the usual result of *Botrytis* infection is a rot most growers prefer to avoid. Normal Botrytis bunch rot often leads to excessive desiccation and rotting of berries before they contain sufficient sugar and subsequent significant loss of tonnage. The rotting berries may also attract fruit flies carrying vinegar bacteria, resulting in a sour-smelling rot, or infected berries may become infected by other undesirable secondary microorganisms.

Symptoms

The first symptoms, in spring, are microscopic: The blossom style becomes infected during flowering and the fungus then remains dormant until later in the season. The bunch rot phase begins with single berries that turn brown and rot, producing visible spore masses, first at skin cracks and then over the entire affected area to produce a characteristic gray mold appearance. Sporulation is enhanced by rain or prolonged irrigation.

Disease Cycle

Botrytis survives the winter by forming a thick, dark resting structure called a sclerotium, either on the surface or within rotted berries and stem clusters. Commonly, these sclerotia are associated with grape mummies left on the vine from the previous harvest.

During moist spring weather the sclerotia produce crops of spores (conidia) that are spread by air currents. There is some evidence that the spores infect the stigmas of the opening grape flowers. These infections then become quiescent and no symptoms are produced on the immature fruit. The factors that keep the fungus in this latent state are unknown. When the sugar content of the grape rises markedly during fruit ripening, the fungus is able to resume growth and spread throughout the berry and into adjoining berries by midseason. Spores are produced on the infected berries—first at cracks and finally over the entire affected area—giving the characteristic gray mold appearance. Cap stems, peduncles and the rachis also may be invaded. Fruit of varieties with tight clusters is rapidly destroyed in this manner.

Spores from infected fruit and other decaying plant material are spread further by air currents and can lead to the next phase: direct infection of intact, ripe berries near harvest. The probability of such infection depends primarily on sugar concentration and on the length of time that relative humidity is

---

*When infection occurs late in the grape's maturity and favorable weather conditions follow, weakening of the grape skin by the *Botrytis* allows water to escape so that the fruits deteriorate into moist shrivelled berries; these partially rotted berries are extremely high in sugar (30 to 40 percent). In certain regions of Europe they are carefully harvested and used to produce sweet, highly aromatic dessert wines that are highly prized. Limited amounts of similar wines have been produced in California's cool coastal districts. "Noble rot" berries of this nature are desirable.

above 92 percent or free moisture is present on the berry surface. The rate of infection also depends on temperature. In general, infections do not progress rapidly at humidities below 90 percent, and in most cases severe bunch rot epidemics are associated with preharvest rains. These storms provide both the moisture and often the optimum temperatures that lie between 15° to 28°C (58° to 82°F) for Botrytis bunch rot development. Infections can dry up or cease development if the weather becomes hot and dry, since *Botrytis* is a relatively cool weather fungus and ceases to grow at 35°C (95°F). (See Table 1.)

TABLE 1. Approximate Moisture Period Required for *Botrytis* Bunch Rot Infection at Various Temperatures.

Temperature	Approximate Moisture Period*
30°C (86°F)	35 hours
26.5°C (80°F)	22
22.5°C (72°F)	15
15.5°C (60°F)	18
10°C (50°F)	30

*Free water (or humidity ≥ 92 percent) at the fruit surfaces, which may be affected by canopy, soil type, and other local microclimatic factors.

Several factors lead to differences in susceptibility among grape varieties. Red varieties contain compounds that inhibit *Botrytis* to some extent, so they are less subject to attack. Tightness of the cluster is also important, since films of water persist longer between tightly packed berries, extending opportunities for rot to develop.

Management of the Disease

Botrytis bunch rot can be best controlled by an integrated program of crop management and chemical treatments. Many loose-clustered or thick-skinned varieties known to be less susceptible may not require chemical treatment.

Good disease management includes:

Sanitation. In areas where Botrytis bunch rot has been a problem, the level of overwintering inoculum should be reduced by removing grape mummies from the vines (at pruning) and discing them under. Soil microorganisms will then be more effective in breaking down the resting sclerotia.

Chemical treatment. In most years, good control of Botrytis bunch rot may be achieved in susceptible varieties by application of a fungicide at bloom time to prevent flower infections. Systemic and contact materials are available. However, the application schedules are slightly different. A systemic fungicide, such as benomyl, should be applied at 1 percent bloom, then 14 days later if bloom continues that long or longer. Benomyl penetrates the unopened flower to provide protection through the open bloom period. Strains of *Botrytis* resistant to benomyl have been isolated in all grape-producing areas in California. The resistant strains do not appear to spread rapidly from vineyard to vineyard, but once established they seem to persist. Growers, therefore, should exercise caution in using benomyl.

In the case of broad-spectrum contact fungicides such as captan, dichloran and maneb, it is recommended that a spray be applied at full bloom (60 to 70 percent caps off) to protect the flower parts from infection, followed by two additional sprays to protect the young berry from infection via adhering senescent flower parts. The full bloom spray is timed to optimize coverage of flower parts that would miss being coated if the flower cap were present.

After fruit softening, fungicide applications are not as effective in controlling Botrytis bunch rot because the clusters have become compact and the spray cannot reach their centers. However, grower experience indicates that in some cases fungus spread may be reduced if the material is applied during or immediately after rain.

Irrigation and canopy management. Moisture is important for the spread of Botrytis bunch rot, and ill-timed irrigations can greatly increase incidence. For disease control vines should not be sprinkler-irrigated once the fruit is mature. If such irrigations are necessary, they should be planned for warm, windy low humidity days to hasten drying. Avoid irrigation in foggy, overcast, humid or cool days to ensure that clusters are not wet more than 15 hours, including irrigation and drying time.

The extent of damage is controlled by microclimate under the vine canopy and the structure of the canopy largely determines the microclimate. In areas where Botrytis bunch rot has been damaging, changes in vine structure (i.e., trellising, canopy density or crop level) may reduce losses.

Leaves take on a tattered, scorched and often misshapen appearance; axillary shoots die and many flowers shell off.

Young canker in cross section appears as a wedge-shaped darkened area coming to a point in the center of the arm or trunk.

Large pruning wound canker on trunk is exposed by removing the bark.

Incipient canker adjacent to pruning wound is revealed by removing bark tissue with a knife.

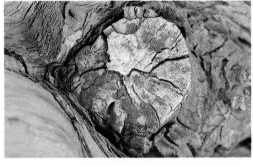

In California's wetter areas perithecia (fruiting bodies) form on diseased wood tissues. Small blackened areas on exposed stub indicate their location.

Mass of fruiting bodies magnified to show numerous small, pimplelike protrusions.

A sharp blade was used to expose spore cavities. Each cavity contains thousands of *Eutypa* spores.

Magnified vertical section reveals a mass of *Eutypa* perithecia embedded in old, diseased grape wood.

Eutypa Dieback 59

misshapen, sometimes cupped, distorted and marginally necrotic with small areas of dead interveinal tissue. Later in the season they take on a tattered and scorched appearance. Many of the flowers shell off; most berries that do establish on these shoots fail to mature. If the shoots are only mildly affected, the tattered leaves appear on just the first few nodes and subsequent growth is normal.

The disease appears first in one or two spurs and spreads in following seasons to adjacent spurs, eventually killing the arm or cordon. Shoots developing from below the affected arm are healthy the first year but show symptoms in subsequent seasons. Unless a major portion of the vine's structural framework is involved, the affected shoots eventually may be covered by normal overgrowth from the vine's healthy portion. It is common to find one side of the vine dead, while the other side appears healthy. When the whole vine has been killed or is severely affected by Eutypa dieback, strong suckers often develop from the still healthy root system. Complete collapse and death of vines or arms in summer is uncommon; once shoots have emerged, they usually grow through summer and die the next winter.

Eutypa dieback disease is not generally visible in vines younger than 5 to 6 years old and is seen most frequently in vineyards established for ten years or more.

An important diagnostic symptom of Eutypa dieback is the formation of pruning wound cankers. These dead areas surrounding large, old pruning wounds often can only be found by removing the rough outer bark. They are frequently located adjacent to the affected spurs. In advanced cases, the wood around an unhealed wound assumes a ridged and flattened appearance so that the trunk or cordon may be twisted and malformed. Older cankers show a marginal zonation, indicating successive annual attempts of the vine to overgrow the necrotic area.

Trunk cankers can be extensive in length and a cross section through the canker often reveals only a narrow strip of live wood. In its early stages, a canker in cross section appears as a wedge-shaped darkened area coming to a point in the center of the arm or trunk. It is characteristic to find streaks or flecks of darkened tissue in the live wood above the canker.

Disease Cycle

Because *E. armeniacae* is a wound parasite, infection invariably occurs through pruning wounds. The fungus has a long incubation period in the vine. Several growing seasons may elapse before visible cankers develop around an infected wound or before stunted shoot symptoms appear. Once an arm or portion of the vine has been killed, it takes several more years before perithecia (spore sacs) are produced on the old infected host tissue and then only under conditions of high moisture. In California perithecia have not been found in the hot, dry San Joaquin Valley, even though the disease is prevalent there, but they are comparatively easy to locate in some of the cooler, more moist areas nearer the coast. There is substantial evidence that viable spores may be carried from the San Francisco Bay Area to infect vines in the San Joaquin Valley's more arid regions. In older diseased vineyards (the Napa Valley, for example) spore-bearing wood is often located on the lower trunk of previously infected dead vines, especially if the remaining "stump" is shaded from direct sunlight. A blackened crust appears as the perithecial layer matures. By using a sharp knife, one can cut through the blackened wood to confirm the perithecial cavities holding the ascospores.

Ascospores are discharged from perithecia during and soon after rainfall; this is the only known means of dispersal and infection. Unfortunately, once the perithecial stage forms it may continue to discharge ascospores for five years or longer. The asexual (*Cytosporina*) spores of *E. armeniacae* do not play any role in the disease cycle.

Recent research suggests that grapevine pruning wounds are more susceptible to infection early in the dormant season than near bud break. One experiment at Davis indicated wound susceptibility of less than two weeks after pruning in February. In apricot trees, wounds usually lose their susceptibility within two to six weeks after pruning, depending on the time of the year. Also for apricots, wound susceptibility decreases faster in late winter than in fall.

Management of the Disease

Because of the diversity of host species which may provide spores for infection, eradication of disease sources is not feasible. However, growers can reduce the risk to their vineyards by eliminating all infected wood of grape, apricot, or other known hosts in the vicinity—especially in the higher rainfall areas.

Late spring is a good time to locate and remove diseased portions of vines before they are masked by vigorous adjacent shoots. Under California conditions, diminishing risk of rain at that time also decreases the chance of spore dispersal and reinfection. However, arm removal has to be done with care, often with a series of successive saw cuts to avoid overcutting. The final saw cut must show completely healthy tissue and no evidence of the pie-shaped sector of dead wood extending downward from the canker site.

Where surgery has been neglected and the vine framework has become extensively involved, it is almost impossible to cut back to healthy wood. In this case, if the vine is weakened to the point that strong suckers are emerging from the lower trunk, the best alternative may be to cut out the dead trunk and rebuild the frame from one of these suckers. Bear in mind that this cutting presents more opportunity for infection, although pruning tools themselves are not known to spread the disease. If possible, make such cuts in dry weather and use a reliable wound protectant. Benomyl paint is registered for this purpose.

In the long term, wound protection offers much better control prospects than eradication, once the disease has become established. Pay special attention to wound protection if drastic retraining or change-over of the variety is contemplated.

EUTYPA DIEBACK CYCLE

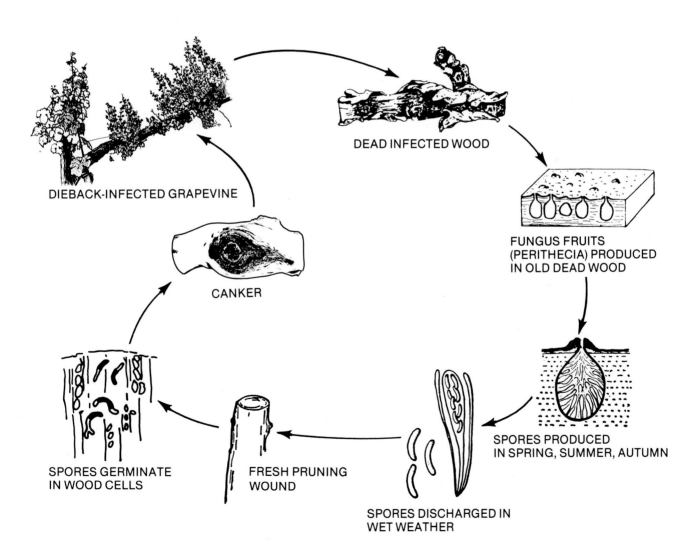

DIEBACK-INFECTED GRAPEVINE

DEAD INFECTED WOOD

FUNGUS FRUITS (PERITHECIA) PRODUCED IN OLD DEAD WOOD

CANKER

SPORES PRODUCED IN SPRING, SUMMER, AUTUMN

SPORES GERMINATE IN WOOD CELLS

FRESH PRUNING WOUND

SPORES DISCHARGED IN WET WEATHER

PIERCE'S DISEASE

Autumn foliar symptoms of Pierce's disease in a red fruit variety, Cabernet Sauvignon.

Petioles of leaves remain attached to cane after leaf fall.

White variety (Chardonnay) with Pierce's disease. "Scorching" of leaves begins at outer leaf margins and progresses inward, beginning in late summer.

Irregular "patchy" bark maturity is prominent on many varieties (Pinot Noir shown here). Fruit may raisin on infected vines.

Early spring symptoms of chronically affected vines are the stunting and distortion of new growth, which emerges later than normal. Note in inset interveinal chlorosis on young leaves.

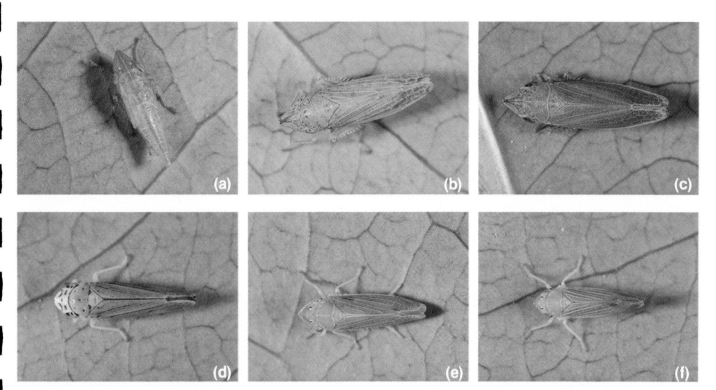

Sharpshooter vectors of Pierce's disease. Green sharpshooter, *Draeculocephala minerva,* is seen here as (a) a nymph; (b) an adult, brown winter phase, and (c) an adult, green summer phase. In photo (d) is pictured the blue-green sharpshooter, *Graphocephala atropunctata,* adult; photo (e) shows the red-headed sharpshooter, *Carneocephala fulgida,* adult female; and photo (f) shows *C. fulgida,* adult male.

Pattern of missing vines caused by Pierce's disease along natural riverbank vegetation, a favorite habitat for the blue-green sharpshooter.

Pierce's disease (PD), a killer of grapevines, is caused by bacteria and spread by certain kinds of leafhoppers known as sharpshooters. It is always present in some California vineyards every year, and the most dramatic disease losses occur in the Napa Valley and in parts of the San Joaquin Valley. During severe epidemics, losses to PD may require major replanting.

Pierce's disease seems to be restricted to portions of North America with mild winters. It has been found in all southern states that raise grapes commercially, from Florida to California, and in Mexico and Central America. In the southeastern states, from Florida through Texas, PD is the single most formidable obstacle to the growing of European-type or *vinifera* grapes.

The disease has been present in California since the 1880s. Known originally as Anaheim disease, mysterious disease, or the California vine disease, it initially broke out in southern California, destroying 35,000 acres. It was extensively studied by N. B. Pierce, and subsequently the disease was named for him. In the Central Valley the disease was first noted about 1917, and from 1933 to 1940 a major outbreak devastated many Central Valley districts. Since then PD periodically has erupted as a serious problem in coastal grape-growing regions and in individual vineyards in the Central Valley. Although average losses most years are small statewide, the losses in individual vineyards can be devastating.

The bacterium that causes Pierce's disease was detected in 1973. U.C. researchers using the electron microscope discovered it in infected grapes. It finally has been cultured, but it has not been scientifically named.

Two factors contributed to the long delay in its discovery. The causal bacteria are limited to the xylem or water-conducting elements of plants and could not be cultivated on the artificial media commonly used to study most plant-pathogenic bacteria. Furthermore, the bacteria went unnoticed in microscope examinations until it was determined that specific staining was necessary to differentiate the organism from the gum in which it is embedded.

This bacterium infects many different plant species in addition to grapevines, some of which show no symptoms of disease. These host plants can serve as reservoirs from which leafhopper vectors can pick up the PD bacterium for transmission to grape. Many host plants also are important food or egg-laying plants for the sharpshooter vectors. Among these are: Bermuda grass (*Cynodon dactylon*), Watergrass (*Echinochloa crusgalli*), Perennial rye (*Lolium perennae*), Fescue grass (*Festuca* spp.), Ripgut brome (*Bromus rigidus*), Blackberry (*Rubus virifolius*), Himalayaberry (*R. procerus*), Willow (*Salix* spp.), Elderberry (*Sambucus* spp.), California mugwort (*Artemisia douglasiana*), Cocklebur (*Xanthium strumarium*), Coyote bush (*Baccharis pilularis*), Stinging nettle (*Urtica holosericea*), Christmasberry or Toyon (*Heteromeles arbutifolia*) and Snowberry (*Symphoricarpos albus*).

Symptoms

Affected vines show symptoms of water stress beginning in midsummer and increasing through fall. The first evidence of PD infection usually is a drying or "scorching" of leaves. The leaf or leaves closest to the point of infection become slightly yellowed (chlorotic) along the margins before drying up, or the outer edge of a leaf may dry suddenly while still green. Typically, the leaf dries progressively over a period of days to weeks, leaving a series of concentric zones of discolored and dead tissue. Red-fruited varieties usually have some red discoloration.

All varieties will have some degree of yellowing on the borders of scorched portions of the leaf. This yellowing may be a very narrow zone in some varieties or include most of the leaf in others. Foliar symptoms gradually spread along the cane from the point of infection out towards the end and more slowly towards the base. Often scorched leaves dry down to the base of the blade and separate, leaving the petiole still attached to the cane. The petiole continues to gradually die back towards its base. The tips of canes on chronically affected vines die back, although this occurs later than the first scorching of leaves.

The woody portions of diseased canes are generally dry, especially on chronically infected vines. The bark on such canes usually matures irregularly. Canes normally maturing tan or brown bark may have mature bark on green canes or islands of green immature bark on matured canes. About the time foliar scorching begins, or slightly later, some or all of the fruit clusters may wilt and dry up or portions of clusters may dry up any time following fruit set. Colored grape varieties may develop color early before wilting and drying.

Usually only a single cane or two will show PD symptoms late in the first season of infection. But in young vines, particularly in sensitive varieties such as Pinot Noir, Pinot Chardonnay or Barbera, symptoms may appear over the entire vine in a single year. In older vines and in some varieties, such as Sylvaner, White Riesling, Petit Sirah or Ruby Cabernet, foliar symptoms usually extend only 10 to 20 inches or less from the point of inoculation during the first season, even in vines infected early in April.

There is variation among grape varieties in the pattern of foliage discoloration and scorching as well as in the speed with which these symptoms appear. Relatively resistant varieties like Sylvaner and Thompson Seedless usually develop extensive yellowing of the leaves, but the marginal drying of leaf tissue is irregular and may occur in patches rather than in concentric bands. In Zinfandel and Mission spots or patches may redden and dry out between major veins in the leaves in addition to marginal scorching.

In general, PD has more severe effects in hot climates than in cooler ones; symptoms appear sooner and vines die more quickly in the Central Valley than in the coastal regions. The time of appearance, extent and severity of symptoms depend to some degree on temperatures and available soil moisture. Shallow soils, moisture stress, or very high temperatures can cause the rather sudden collapse of vines or portions of vines.

Chronically-affected vines are slow to begin growth in the spring. On canes or arms that had foliar symptoms the preceding fall, new growth will be delayed up to two weeks and will be somewhat dwarfed or stunted. Some canes or spurs may fail to bud out at all. Leaves on stunted shoots may also have a faint yellow mottling between major leaf veins and may also be distorted or asymmetrical. Shoot wilting has been observed on rare occasions. If fall symptoms are confined to one arm or cordon branch of the vine, usually only that part of the vine will have symptoms the following spring. In cases where most of the vine has PD symptoms, especially if many arms or canes are dead, suckers may sprout from the base of the trunk or rootstock. *However, except when severely infected, most vines that show stunted early growth because of PD will produce near normal growth from late April or May through late summer, at which time leaf burning reappears.*

The symptoms of several other disorders of grapes can be confused with PD. (Some of these are discussed elsewhere in this book.) Any disorder that weakens or causes deterioration in vine growth in the late summer or fall may lead to delayed or stunted growth the following spring. Zinc deficiency will produce small or stunted leaves with interveinal mottling similar to that of PD. Phylloxera causes a decline of vines. Oak root fungus (*Armillaria* root rot) can cause wilting along with discoloration and drying of the fruit and foliage. Some of the symptoms of Eutypa dieback and of measles are similar to those of PD. Leaf scorching caused by excessive salts in the soil can sometimes be confused with that caused by Pierce's disease. Misapplied herbicides also can cause leaf chlorosis or scorching.

Vectors

The green and red-headed sharpshooters are the primary vectors in California's Central Valley. The blue-green sharpshooter is the most important in such coastal regions as the Napa Valley. All three species are found in both coastal and central valley areas. Differences in the patterns of PD spread within these two major geographical areas are largely because of seasonal differences in population levels of the vectors.

Central Valley. The greatest concentration of Pierce's disease in the Central Valley occurs near habitats favorable to the green sharpshooter or the red-headed sharpshooter. The greatest amount of disease spread is usually downwind from pastures, weedy hay fields or other vector-source areas. A relatively small area of grasses along a roadside or irrigation ditch is sufficient for either the green sharpshooter or red-headed sharpshooter to reach damaging levels. They can acquire PD bacteria from symptomless plant hosts as well as from diseased grape or alfalfa.

In the Central Valley, the green and the red-headed sharpshooters are chiefly grass feeders; they are only rarely found feeding on grape. The green sharpshooter reaches its peak abundance in permanent pastures, weedy alfalfa fields or in places where water stands for prolonged periods and promotes the luxuriant growth of grasses. Watergrass and Bermuda grass seem especially attractive, although this insect also reaches high densities on Italian rye, perennial rye or fescue. The green sharpshooter is relatively slow to move into new sites and usually requires more than one year to build up dense populations. It is often abundant in permanent orchard cover crops or localities with perennial weed cover. It is not abundant in orchards or vineyards that are thoroughly clean-cultivated at least once a year, even when such plantings have dense grass cover during the remainder of the year.

The green sharpshooter typically has three generations per year in the Central Valley. It is active as an adult in the winter. Females insert eggs into the leaves of winter annual or perennial grasses beginning in February and March, and nymphs emerge from late February through March. Second generation eggs are laid beginning in April or early May. Nymphs from the second generation reach maturity during the latter part of June through July, during which time third generation eggs are deposited. Adult coloration can vary from bright green through dull brown during winter, whereas during late spring and summer all adults are a bright grass-green color.

The red-headed sharpshooter is not as common or as abundant as the green sharpshooter. It prefers more open or sparser plant growth than does the green sharpshooter and is most often found on Bermuda grass. Grapes are not a preferred host. There are usually four generations per year. Eggs of each generation are laid in mid-March, mid-May, early July and mid-August. Approximately 25 days are required for development from egg to adult. Adults are active during winter, but they are much less abundant and more widely scattered than the green sharpshooter adults are during this time. Movements of the red-headed sharpshooter are thought to be similar to those of the green sharpshooter, and adult populations may disperse rapidly from habitats which dry up and become unsuitable for feeding. The red-headed sharpshooter may be found in small patches of Bermuda grass along roadsides, ditches or the margins of alfalfa fields where the growth of grass is not succulent or dense enough to support high populations of the green sharpshooter.

Coastal areas. The blue-green sharpshooter, *Graphocephala atropunctata*, unlike the grass-feeding sharpshooters, feeds, reproduces and is often abundant on cultivated grape. It will feed and reproduce on many plants, but it prefers woody or perennial plants such as wild grape, blackberry, elderberry, mugwort and stinging nettle. The blue-green sharpshooter is most common along stream banks or in ravines or canyons that have dense growth of trees, vines and shrubs. It is sometimes abundant in ornamentals around homes on such plants as rose, fuschia, citrus, ivy and geranium. It is seldom found in unshaded, dry locations because it feeds on succulent new growth in areas of abundant soil moisture and shade.

The blue-green sharpshooter has one generation per year in most of California, with a second generation in some parts of the state. In late winter and early spring adults become active in short flights in natural vegetation along streams or canyons. They begin moving into nearby vineyards for feeding and egg laying, once grape shoots are several inches long. Their dispersal into vineyards increases as the natural vegetation begins to dry up. Most overwintered adults die out by the end of June. Nymphs from their eggs emerge in May through July. Some of these become adults by late June, and the number of young adults continues to increase through July and August. In August, when grape foliage is less succulent, sharpshooters begin to move back into nearby natural habitats.

Management of the Disease

Most dispersal of Pierce's disease seems to develop directly from outside source areas to individual grapevines within vineyards. There is no appreciable secondary spread or vine-to-vine infection by sharpshooters or other means. There also is no evidence that PD can be spread by contaminated pruning tools. Furthermore, many studies in both coastal and interior areas have shown that rigorous removal of infected vines is not beneficial in reducing the spread of PD within a vineyard, although removal and re-planting may be necessary to keep up vineyard productivity. Therefore, the only way to reduce or prevent PD spread through vector control is to prevent sharpshooters from entering vineyards.

Central Valley. In the Central Valley control means eliminating or preventing the conditions conducive to buildups of green and red-headed sharpshooters. Insecticide treatments are only temporarily effective against adults and of little value overall

Large numbers of spots coalesce to give a scabby appearance to the base of shoots.

Leaf spots (usually on basal leaves) show up as tiny black spots with yellowish margins.

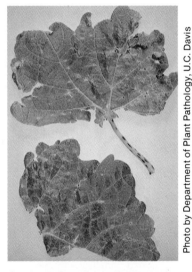

Severe infection may kill or distort portions of the leaf blade and cause early leaves to fall because of petiole necrosis.

Photo by Department of Plant Pathology, U.C. Davis

Photo by Department of Plant Pathology, U.C. Davis

Spore-bearing bodies of *Phomopsis* appear on this infected berry.

Heavily infected shoots near the head of this Thompson Seedless vine are retarded or killed, resulting in a bare center which reduces available wood for the following season.

This disease, caused by the fungus *Phomopsis viticola*, results in small spots or lesions on grape leaves, shoots and cluster stems. Injury reduces yield.

Phomopsis disease was first seen in 1935 in the American River section of Sacramento County. Originally, reports and descriptions earlier in the century—especially for the northeastern United States—mistakenly included dead arms as a symptom. Thus, for many years, *Phomopsis* cane and leaf spot was known as "deadarm." Research at Davis helped to establish that deadarms and pruning wound cankers on vines are caused not by *P. viticola* but by another fungus, *Eutypa armeniacae*. (The disease caused by *Eutypa* is discussed elsewhere in detail.)

In wet spring years, *Phomopsis* shoot and leaf necrosis can be particularly severe on Tokay, Thompson Seedless and Grenache varieties growing in the Central Valley between Lodi and Visalia. Other varieties grown in this area susceptible to a lesser degree include Cardinal, White Malaga, Emperor, Calmeria and Rish Baba. In districts other than the Central Valley, *Phomopsis* has been seen on other varieties, but it is not economically important.

Actual economic loss from *Phomopsis* in most years is minor. However, in years of severe infection, losses can occur from:

• Shoots breaking off near the base where heavy lesions develop, reducing the cluster count and yield.

• Retention of infected weakened wood at pruning time, resulting in reduction of the next season's crop level. Early symptoms may be difficult to see, and selecting good, clean, mature wood is time consuming and can increase pruning costs.

• Fruit infection, which occasionally occurs before harvest under cool, wet weather conditions.

• Reduced storage capability; infected fruit stored under cool temperatures and high humidity can develop *Phomopsis* rot.

Symptoms

Infection occurs on leaves, shoots, cluster stems and berries.

Leaves. The first symptoms that appear on leaf blades and veins are tiny dark brown to black spots with yellowish margins. These spots first show three to four weeks following a rain after budbreak. If large numbers develop, they may kill portions of the leaf. Basal leaves with heavy infection become distorted and usually never develop to full size. When leaf stems are heavily infected, the leaves turn yellow and abscise. Later, normal leaves develop on subsequent nodes, hiding the distorted basal leaves.

Shoots. Small spots with black centers, similar to those found on the leaves, are the first evidence of shoot infection. This infection usually occurs on the basal portion of the shoots. When these oblong spots become a few millimeters in length, the epidermal layers of the shoots usually crack at the infected parts. Where the spots are in large numbers, they coalesce and may ultimately give a scabby appearance to parts of the shoots. Heavily infected shoots can be dwarfed or retarded and some may die. Later, after shoots have developed 300 to 600 mm (12 to 24 inches) long, shoot breakage can occur during strong winds. This breaking occurs in tissue areas of heavy scabbing and lesion development, usually below the clusters. Pathologically, shoot lesions become inactive during the summer.

Clusters. During spring spots similar to those on the shoots and leaves also appear on the flower cluster stems. Occasionally the cluster stems are so badly infected that the clusters wither. These lesions also become inactive in summer, but early fall rains combined with cool weather may reactivate the disease, resulting in berry and bunch rotting. Fruit symptoms are generally not extensive, with only isolated bunches affected on any one vine. However, rain just before harvest can cause light brown spots on clean berries; the spots enlarge quickly and become dark brown. Black pycnidia or spore-bearing bodies, often in concentric rings, break through the skin and yellow spore masses may exude. Finally, the berries shrivel and become mummified.

Infected canes. During the late dormant season, infected wood areas on basal portions of the canes appear bleached. Severely affected canes or spurs exhibit an irregular, dark brown to black discoloration intermixed with the whitish, bleached areas. Tissue in the vicinity of the original lesions and at the nodes is also whitish with black speckling.

These black specks are the pycnidia which develop during the dormant season, break through the surface and appear as minute, black, pimplelike pustules. They are the source of overwintering spores for the following season. Severely affected canes and spurs are more sensitive than healthy tissue to low temperatures that can cause extensive killing of the spurs and weakening of the canes.

Disease Cycle

Infection generally occurs in the spring when the shoots begin to grow. Rain is required for infection. Around budbreak spores released in large quantities from the overwintering pycnidia on diseased canes, spurs and bark are splashed by rain onto newly developing shoots. Infection occurs when free moisture remains on the unprotected green tissue for many hours; symptoms become visible shortly afterwards. Obviously, heavy and prolonged rains in late March and April, soon after budbreak, are ideal for spring infection. But the number of basal nodes affected will vary according to the weather suitable for infection. Hence, the disease will vary in occurrence and severity from season to season.

Management of the Disease

Because spur and cane lesions provide most of the inoculum for new infections, reducing this source of the disease is important; carefully prune out badly infected canes to reduce the carryover of spores.

Use an eradicant chemical late in the dormant period (at least one month after pruning and tying the canes if sodium arsenite* is used) to help clean up overwintering inoculum and lessen the risk of new shoot infection. Both sodium arsenite and dinoseb* (Premerge) (2-sec-butyl-4,6-dinitrophenol) have been successfully used. Thorough application of one of these eradicant materials during late dormancy will insure against shoot infection if early spring turns unusually cool and wet. However, there are disadvantages. Both chemicals require safety measures. They can also be toxic to vines if used with improper timing. Premerge can damage buds that have begun to swell and should only be used during complete dormancy. On the other hand, sodium arsenite can be used through budswell to bud break.

Growers should check with their grape buyers before applying sodium arsenite. Not all wineries approve its use, although it is legally approved by regulatory agencies at the time of writing.

Another alternative is to treat vines with a fungicide during the early shoot growth stages. Captan, folpet and mancozeb all give satisfactory protection of young shoots if applied *before* prolonged cool, wet weather. Sprays applied from bud break to 1.3 cm (½ inch) shoot length and possibly again when shoots are 125 to 150 mm (5 to 6 inches) long will provide good control. An additional foliar spray may be necessary after heavy spring rainfall.

In severely affected vineyards, both dormant and spring treatments may be advisable.

PHOMOPSIS CANE AND LEAF SPOT CYCLE

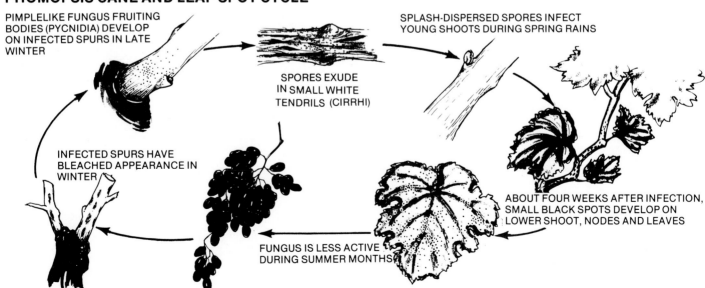

PIMPLELIKE FUNGUS FRUITING BODIES (PYCNIDIA) DEVELOP ON INFECTED SPURS IN LATE WINTER

SPORES EXUDE IN SMALL WHITE TENDRILS (CIRRHI)

SPLASH-DISPERSED SPORES INFECT YOUNG SHOOTS DURING SPRING RAINS

INFECTED SPURS HAVE BLEACHED APPEARANCE IN WINTER

FUNGUS IS LESS ACTIVE DURING SUMMER MONTHS

ABOUT FOUR WEEKS AFTER INFECTION, SMALL BLACK SPOTS DEVELOP ON LOWER SHOOT, NODES AND LEAVES

*Restricted material; permit required from County Agricultural Commissioner for possession or use.

MEASLES

Striking leaf symptoms are most prevalent in July and August. Entire vines or only portions may be affected.

Early leaf symptoms appear as small chlorotic interveinal areas.

Yellow margins surround the dead interveinal areas as measles progresses.

Severely affected leaf shows death of most of the leaf blade.

74   Measles

Some berries have cracked and shrivelled in a bunch affected by measles.

The peculiar berry spotting gives measles its name. The berry skin is peppered with small, round, dark spots.

Berries dry up or rot on vines severely affected.

Sometimes called "Black Measles," "Spanish Measles," or even "Apoplexy," measles disease was first described on grapes in France, where it is known as "Esca." Measles disease can be found on wine, table or raisin grapes in most areas of California, although it is more prevalent in areas with consistently high summer temperatures, such as the Central Valley. Generally, plantings that are 10 years of age or older are the ones affected, although it has been seen on fruit and foliage in younger vines. Crop loss estimates are difficult to make since the disease occurs seasonally; a vineyard may show variable symptoms from season to season. Table grape growers, especially those with older Thompson Seedless, Red Malaga or Emperor vineyards in the San Joaquin Valley, suffer the most serious fruit losses. Affected clusters are not salable because of their disagreeable appearance and flavor; they are also more susceptible to bunch rot.

The cause of measles is not fully known, although appearance of symptoms is invariably correlated with an internal wood rot, beginning at large pruning wounds. Species of fungi in the genera *Fomes, Cephalosporium* and *Stereum (Phellinus)* are most frequently mentioned in the literature as being associated with the disease.

Symptoms

The most important symptom to the grape grower is the peculiar berry spotting, most noticeable on white varieties. The skin is peppered with small, round dark spots, each bordered by a brown-purple ring; these spots may appear any time between fruit set and ripening and affect either entire or parts of clusters. In severely affected vines the berries often crack and dry on the vine or are subject to spoilage.

Leaf symptoms usually develop on canes with measled fruit, but also on canes with normal fruit. Typically, affected leaves display small, chlorotic interveinal areas which enlarge and dry out; in dark colored varieties, dark red margins surround the dead interveinal areas. Severely affected leaves drop and the canes die back from the tips. Symptoms may appear at any time during the growing season, but they are most prevalent during July and August. Entire vines or only portions of a vine may be affected; occasionally the symptoms appear so quickly and dramatically that the vine dies.

Disease Cycle

As previously indicated, the real cause of measles is not known, although it is assumed to be wood-rotting fungus spores that gain entry through large pruning wounds. As the wood rot develops over the years, it is suspected that toxins produced by the fungal microorganisms are transported to portions of the vine and result in measles symptoms. Measles occur sporadically and may show in vines during one season but not during the next, although badly affected vines usually exhibit symptoms every year. Old, dead, rotting trunks of grapevines probably provide a suitable habitat for production of the fungus fruiting bodies that give rise to spores of the measles causal agent.

Management of the Disease

Dormant treatment of affected vines with sodium arsenite* has been reported to control measles if the solution is applied to the entire surface of the trunk and arms, thoroughly wetting old pruning wounds. However, control is often erratic, and there is some risk involved both to the applicator and the vines. Grapevines should be pruned before spraying to improve coverage and reduce the amount of chemical needed. The spray application should be made at least four or more weeks after pruning. This length of time is needed to permit adequate healing of leaf scars and pruning wounds, so that the chemical will not be absorbed to systemically damage adjacent buds. Applications can be made up to bud break.

Sodium arsenite is hazardous to the applicator and requires certain safety measures in application. Growers should also check with their grape buyers before using it. Not all wineries approve of its use, although it is legally approved by regulatory agencies at the time of writing.

---

*Restricted material; permit required from County Agricultural Commissioner for possession or use.

Long known to occur near old creek beds and water courses where oaks once grew, *Armillaria* root rot (oak root fungus) is found on a wide range of woody plants in California. It is of limited economic importance in statewide grape production, except where broad-scale soil fumigation is required during vineyard preparation. When it occurs, small patches of vines are subject to a decline and collapse. Preplant precautions for control are sometimes necessary—especially in vineyards in the Napa, Sonoma, Santa Clara, Salinas and northern San Joaquin valleys. It is often found in vineyards that have replaced old orchards of prunes, walnuts or other susceptible tree crops that allow the fungus to build up. However, the causal fungus, *Armillaria mellea*, rarely affects many acres of vineyards. More commonly, it results in losses of a few vines at a time.

Symptoms

Affected vines decline in vigor, the leaves take on a yellow color in early summer, and cane growth becomes weaker before the vine eventually collapses. Patches of diseased vines may gradually increase in size each year if left untreated.

Oak root fungus causes a moist rot of the crown and root tissue, beginning just below ground level. To check a suspected infection, remove a spadeful of dirt from beside the trunk and, with a knife, cut into the bark below ground level. The presence of flat, creamy-white plaques or fans of fungus mycelium with a distinct mushroom odor beneath the bark help confirm field diagnosis. The white fungus growth never appears on the outside of the bark, although rhizomorphs may occasionally be found. These are dark brown, shiny, rootlike strands that enable the fungus to move from one part of the plant to another. Another sign of infection is the appearance of honey-colored mushrooms around the vine base after early fall rains.

Disease Cycle

The fungus survives on diseased wood and roots below ground for many years. Healthy plant roots can become infected when they come in contact with this inoculum from a preceding orchard crop or nearby oak trees. Flood waters (also deep tillage equipment) sometimes help spread infected roots in a vineyard; the fungus is favored by soil that is continually damp. The mushrooms are not considered significant in disease spread.

Management of the Disease

Because there are no known *Armillaria*-tolerant grape rootstocks, preplant chemical fumigation of the soil is the only control for oak root fungus. Treatment is best undertaken in September to November when the soil is still dry. Several preparatory steps are involved:

• Before planting or replanting an affected soil, all diseased vines, tree stumps and roots greater than 36 mm (1½ inches) in diameter should be removed, piled and burned.

• In treating portions of an existing vineyard, healthy appearing vines adjacent to those showing symptoms often are also infected and should be removed. If taken out, include their soil area in the fumigation treatment.

• Dry out the soil as much as possible. The drier the soil, the deeper the chemical will penetrate and the more effective the treatment will be. Do this by withholding water during the summer and by using cover crops (such as sudan grass or safflower) to further deplete soil moisture. Finally, deep till the dried area, being careful not to spread any diseased roots.

Two materials are available for fumigation: carbon bisulfide* and methyl bromide.* Both are dangerous materials and must be used with care. Methyl bromide achieves greater penetration, but the soil must be tarped for at least two weeks to ensure adequate retention of the gas near the field surface. The treated soil must be aerated for at least a month before planting to make sure no gas remains to damage new vines.

*Restricted material; permit required from County Agricultural Commissioner for possession or use.

Small areas can be effectively hand injected, but commonly growers call in a local contractor to do the job with specialized machinery. Soil fumigants can be hazardous, and these restricted-use pesticides should be used carefully.

It should be pointed out that soil fumigation provides a period of protection generally lasting from six months to six years, depending on factors such as rooting depth and soil conditions at time of application; rarely does it eradicate oak root fungus, and retreatment may be necessary in localized areas. If the soil is treated when wet or if it has extensive clay or silt layering in the rooting zone, it is doubtful that fumigation will be successful.

Note: The dry soil treatment is specific for deep dwelling soil organisms, such as oak root fungus and nematodes, whereas moist soil is helpful for successful fumigation to control weed seeds and fungi that inhabit soil surface layers.

Armillaria-affected vines often collapse in midseason.

Exposure of the crown area reveals white fungus fans beneath the bark. Infected bark tissues are soft and moist; fungus growth has a strong mushroom odor.

Photo by William J. Moller

Armillaria fruiting bodies.

Photo by William J. Moller

Armillaria mycelium growing on wood tissues.

Photo by William J. Moller

Swollen, knobby outgrowths or galls often found at the base of the vine.

Photo by Austin C. Goheen

Freeze-damaged tissues can result in galls along the length of the cane.

Photo by Austin C. Goheen

Gall tissue may sometimes extend up the vine trunk.

Photo by Amand N. Kasimatis

Plant infected by crown gall in the nursery.

Until recently crown gall has not been a problem in California vineyards. Today it is of increasing incidence as a result of distribution of contaminated planting material produced under mist and because of the more favorable disease climate associated with the increasing use of sprinkler and drip irrigation in vineyards. Many crown gall-affected plants show only minimal symptoms while others become unthrifty and some may die. Older vines were thought to be relatively tolerant of infection, but recent unpublished studies indicate that large galls on the vine base can cause a significant decrease in yield.

Crown gall, also known as black knot when it occurs on the aerial parts of grapevines, is caused by the bacterium *Agrobacterium tumefaciens*. Many woody plants are susceptible to it, and the disease often results in serious damage to deciduous fruit trees, bush berries, shrubs and rose bushes.

There are three types of *A. tumefaciens* occurring in California. The first two types have broad host ranges including grape, whereas the third type infects only grapes, as far as is known. Grape plantings in recent years have been infected primarily with the third type. (There is interest today in another *Agrobacterium* species, *A. radiobacter*, because one strain has been used as a biological control of crown gall on deciduous fruit trees and some ornamentals. *A. radiobacter* differs from *A. tumefaciens* only in that it cannot cause disease. Preliminary tests indicate that it will not be effective against the grape strain of *A. tumefaciens*.)

Symptoms

Gall formation, the typical symptom of the disease, is first seen as a small outgrowth on roots, crowns and, sometimes, on canes and stem tissue. The surfaces of young galls are smooth; they become rough as the gall ages and increases in size. Galls may grow to be several inches in diameter and are composed of soft, disorganized tissue. They lack the rings of bark, cambium and wood found in healthy tissues or callus. Old galls become dark, brittle and cracked. Galls above ground level usually dry out and break off in dry climates, and root galls, if left exposed, often rot and slough away with time, especially during hot, dry summer weather.

Disease Cycle

The pathogen can be transmitted by any agent that contacts contaminated material. The bacterium must enter the plant through a wound; thus, any injury is a potential site of infection. Galls commonly develop where plants have been suckered or injured during cultivation or pruning. Galls frequently will appear where the vine tissue has been damaged by a severe freeze. Natural growth cracks in woody tissue also appear to be good sites for infection.

Nursery operations that produce wounds, such as making cuttings for rooting, grafting, transplanting, root trimming and budding, offer many opportunities for *A. tumefaciens* to invade. Mist propagation in nurseries provides excellent conditions for infection and development of crown gall and is an especially serious problem.

Management of the Disease

Crown gall may be controlled in several ways, the most important being sanitation and avoidance of injuries. In greenhouses and nurseries, clean equipment, work areas and soil are important, as well as careful handling of plants. Although experimental evidence shows that soil fumigation does not eradicate the organism, some nurserymen feel they have been helped by the practice.

Photo by Mary Ann Sall

Affected vines appear weak and stunted and usually develop premature fall colors. Beware of confusion with such other root disorders as corky bark virus and Armillaria root rot.

Photo by Armand N. Kasimatis

If soil is removed from the crown of the vine, darkened bark and wood tissues extending to the roots will be revealed by cutting with a knife.

Phytophthora crown and root rot, sometimes known as collar rot, is a minor disease of grapevines that can occur anywhere in California. Occasionally seen on young vines, especially under drip irrigation, crown rot is favored by an excess of soil moisture around the base of the plant. The water mold fungus *Phytophthora* is responsible for the disease.

Symptoms

Vines appear stunted and develop premature autumn colors during the growing season. Removal of soil around the base of the plant and examination with a knife reveals a canker around the trunk extending downward to the roots, which become blackened and decayed. There is no white fungus growth, as found with Armillaria root rot. The necrotic bark usually occurs at or near the soil level, effectively girdling the vine.

Disease Cycle

Excessively wet soil at the crown of the plant and in the root zone contributes to disease development. The fungus is soilborne, and prolonged and repeated occurrence of free water in the vineyard following rain or irrigation will encourage crown rot.

Management of the Disease

Try to eliminate free water and prolonged soil saturation around the base of vines by careful soil water management; remove weeds which prolong high humidity. Where drip irrigation is used, locate emitters a foot or so away from the vine trunks, and if vines are on a rootstock, be sure to keep the graft union above the soil surface. No chemical control measures are presently recommended for crown and root rot on infected vines.

Symptoms

Leaves begin to wilt and collapse in early summer heat, followed by death of some shoots, vascular discoloration and streaking of wood. Frequently vines are only partially affected, and strong new growth often appears in unaffected portions. Wilted leaves normally remain attached, and fruit clusters at the base of affected canes dry up. Vines that are not killed may show complete recovery by the following year, as is the case for certain other woody plants.

Disease Cycle

The fungus survives in the soil and on other plant hosts such as weeds, vegetables and cotton. These plants can build up a reservoir of dark, microscopic resting spores of *Verticillium*. In other woody hosts, root infection is favored by cool spring weather and high soil moisture. Once established in the grape roots, the fungus migrates up the vascular tissue, and wilt symptoms appear when warmer weather begins. Exposure to prolonged hot summer temperatures eradicates the fungus from the tops of certain tree crops; this probably also occurs with grapevines and would explain the recovery of badly affected plants.

Management of the Disease

Verticillium wilt has not occurred frequently enough in California vineyards to warrant control measures. Avoiding other susceptible crops for three years before planting is the safest precaution against losses.

Photo by William J. Moller

Wilt and collapse frequently occur only on one part of the vine.

Verticillium wilt causes significant losses in several annual and perennial crops in California, but it is a minor disease of grapes. Caused by the fungus *Verticillium dahliae*, it first was found in 1973. Disease incidence as high as 15 percent in certain young vineyards was reported at that time. But in general, incidence is very low (1 to 2 percent) and may be seasonal in occurrence. It has been found only in new vineyards planted on sites previously used for wilt-susceptible crops, such as tomatoes, melons or other vegetables.

Photo by Austin C. Goheen

Leaves lose turgor and show signs of scald because of invasion of xylem tissues by *Verticillium*.

Photo by William J. Moller

Verticillium-affected canes exhibit vascular streaking.

Photo by William J. Moller

Water-conducting tissues of severely affected canes may be blackened by the fungus.

Photo by William J. Moller

Berries on affected shoots dry up early in season.

The viruses and similar agents that cause important diseases in grapevines may be difficult or even impossible to isolate for identification, but the diseases they cause have one common characteristic: Each affects all parts of the host plant and consequently spreads with the cuttings or other propagating materials taken from the infected mother plant. New vineyards set from such cuttings or budwood will consequently be infected from the start, with losses varying, depending on the effect of the individual virus. Substantial damage can result because infected plants do not spontaneously recover from virus infections but remain infected for life. Therefore, to prevent losses to virus, it is necessary to recognize the diseases and to avoid using affected mother vines in plant propagation.

In California we recognize a number of virus diseases of grapes, but are concerned principally with only three: leafroll, corky bark and fanleaf degeneration. Other diseases are restricted to local occurrences or cause no measurable reduction in yield, quality or vigor. (Pierce's disease [PD], once thought to be caused by a virus, is actually caused by a bacterium and is discussed in a separate section.) Growers, managers and advisers should learn to recognize the symptoms of the three important grape virus diseases because there are no visible pathogens.

—————— LEAFROLL ——————

Leafroll disease occurs wherever grapes are produced. It is especially common in California in vines grafted to rootstocks for phylloxera or nematode control. It can also occur in own-rooted vines, if such plants are propagated as cuttings from affected mother vines. It has been found in many new grape varieties and advanced selections from grape breeding programs, where vines have been grafted on affected rootstocks during the testing period before release. In some varieties all plants became infected and no leafroll-free vines were available until heat treatments for the elimination of disease were developed.

Loss from leafroll in a single year is not catastrophic, but when measured year after year over the life of a vineyard, it is considerable. When the perennial nature of the disease and its worldwide distribution are considered, it is undoubtedly one of the most important contributors to production losses. For example, our studies have shown yields on affected vines have been reduced by 20 percent and fruit maturity delayed by three weeks to one month. The loss from reduced yield is obvious; but that from delayed maturity is more subtle, depending somewhat on (1) projected uses of the grapes, such as for table fruit, raisins or wine must, (2) the advent of fall rains and (3) the length of the growing season.

Symptoms

Leaf and fruit symptoms can be used to diagnose the disease in many vinifera cultivars. Leaf symptoms develop in early summer in vineyards stressed by lack of water, then somewhat later in those irrigated. Symptoms become distinct in both the leaves and fruit of most vinifera varieties as the crop matures. The strongest leaf symptoms appear between the time of harvest and leaf fall. On affected vines, the margins of the leaf blades roll downward, starting with the basal leaf on the cane. Areas between the major veins turn yellow or red, depending on whether the variety produces green- or red-colored fruit. In either case, the area immediately adjacent to the major veins remains green. The most characteristic symptom is rolled yellow or red leaves with green vein-banding. The entire vine takes on a yellowish or reddish cast at harvest time and on into autumn. In a few cultivars the red or yellow areas of the blade continue to degenerate so that the blade becomes necrotic as if it were burned. There is an exception in certain table grape cultivars, such as Thompson Seedless, where leaf burning *without* rolling is the principal symptom. In all varieties at harvest fruit maturity is delayed so that fruit on the affected vine is still greenish or whitish when fruit on healthy vines is ripe. (This symptom led to the name White Emperor disease, which leafroll was once called in Emperor table grape vineyards in California.)

Symptoms on affected vines are indistinct during dormancy and the early part of the growing season. The vines may be slightly smaller than healthy ones, but this, too, is not easy to see.

Leaves on red grape varieties affected by leafroll assume a vivid dark red between harvest and leaf fall. Areas between major veins turn red (see inset), if the variety produces red colored fruit, and margins of the leaf blade roll downward. Pinot Noir is pictured here.

American rootstock varieties do not show symptoms at any time during the season, making such vines impossible to identify from symptoms in the field.

Spread

Leafroll spreads slowly from vine to vine in California vineyards, yet the incidence of the disease is often nearly 100 percent. This high incidence results from its semi-latent nature and from methods used in propagating new vineyards, especially those grafted onto clonally propagated rootstocks. The disease has little effect on survival of either buds or cuttings, so it can be introduced to new plantings from either the scion or rootstock source or both. If these are not carefully selected for freedom from disease—and this condition is often hard to determine when gathering buds and cuttings—leafroll will increase. This comes about purely by chance because both combinations, diseased buds on healthy rootstocks and healthy buds on diseased rootstocks, produce diseased plants; healthy plants can be produced only by healthy buds on healthy rootstocks. Within a relatively few propagations without selection of wood sources for disease freedom, the disease incidence approaches 100 percent.

Management of the Disease

Control the spread of leafroll or any other virus disease by using healthy mother plant materials. See section *Management of Grape Virus Diseases,* page 92.

White grape varieties show yellowing and leaf rolling. Diseased Chardonnay vine is left, healthy right.

The most characteristic symptom on white-fruited varieties, rolled leaves with green veinbanding, is seen on this Chardonnay leaf.

Yield can be markedly reduced; diseased bunch of Emperor table grape is right.

At harvest fruit maturity is delayed so that fruit on an affected Cabernet Sauvignon vine is still immature (as shown by hand held bunch), while fruit on healthy vine is ripe.

Damage sometimes confused with leafroll is caused by the three-cornered alfalfa hopper *(Spissistilus festinus)*. Inset shows hopper and girdle injury on leaf petiole.

─────── CORKY BARK ───────

Photo by Austin C. Goheen

Corky bark-affected vines on St. George rootstocks often die back, leaving only the rootstock where the vine should still be.

Corky bark occurs worldwide in much the same pattern as leafroll, but it is generally less common. It has become a widespread and serious problem in those countries where nurserymen do not pay close attention to wood selection. It may be present in a latent state in many vinifera cultivars, and symptoms do not show until affected buds are grafted onto phylloxera-resistant rootstocks. When this happens, a fairly normal appearing plant may grow initially but incompatibility gradually develops at the union. This causes the scion shoot to gradually decline and ultimately die.

In California the disease is most common in old vineyards grafted on St. George rootstocks.

Photo by Austin C. Goheen

Photo by Jeff Hall

In late autumn leaves on corky bark vines do not drop normally and some remain attached to the vine for several weeks after frost. Unlike leafroll, the leaf color spreads uniformly across the entire leaf blade. In the inset is seen a comparison of healthy (left), leafroll (center) and corky bark-affected (right) leaves. Corky bark leaves roll downward and turn uniformly reddish-yellow.

Corky bark-affected vines can be detected during dormancy by the proliferation of the rootstock.

If an affected rootstock is cut transversely in the below-ground-level portion, the woody cylinder is deeply convoluted and the inner wood is stained pink.

GLOSSARY

Grape Diseases

Ascomycetes—One of the five classes of fungi; contains many plant pathogens.

Ascospore—A spore resulting from meiosis and borne in an ascus.

Ascus—A microscopic, sac-like structure generally containing eight ascospores; characteristic of the ascomycetes.

Bacterium (pl. *bacteria*)—Microscopic one-celled organism that reproduces by fission.

Blight—Common name for a number of different diseases on plants, especially when collapse is sudden, e.g., leaf blight, blossom blight, shoot blight.

Canker—A lesion on a stem.

Chlamydospore—Thick-walled, asexual resting spore.

Cleistothecium (pl. *cleistothecia*)—A perithecium without a special opening; in powdery mildews.

Conidium (pl. *conidia*)—Any asexual spore except sporangiospore or chlamydospore.

Eradicant—A chemical used to eliminate a pathogen from the host or environment.

Eradication—Control of disease by eliminating the pathogen after it is already established.

Fumigant—A substance that forms vapors that destroy pathogens, insects, etc.

Fungus (pl. *fungi*)—An organism with no chlorophyll, usually reproducing by spores; mycelium with well-defined nuclei.

Gall—Outgrowth or swelling of unorganized plant cells produced as a result of attack by bacteria, fungi or other organisms.

Haustorium (pl. *haustoria*)—A specialized outgrowth of a stem, root or mycelium (in fungi) that penetrates the host plant and absorbs food.

Host—Any plant attacked by a parasite.

Hydrophobic—Disliking water.

Infection—Process of beginning or producing disease.

Inoculum—Pathogen or that part of it that infects plants (e.g., spores, mycelium, etc.)

Lesion—Localized spot of diseased tissue.

Mildew—Fungus growth on a surface.

Mold—Fungus with conspicuous mycelium or spore masses; often saprophytic.

Mycelium (pl. *mycelia*)—Mass of fungus hyphae.

Parasite—An organism that lives on or in a second organism, usually causing disease in the latter.

Pathogen—Any disease-producing organism.

Perithecium (pl. *perithecia*)—Flasklike fungus fruiting structure of some ascomycetes that contains asci and ascospores.

Protectant—A chemical applied to the plant surface in advance of the pathogen to prevent infection.

Pycnidium (pl. *pycnidia*)—Flasklike fungus fruiting structure of some Fungi Imperfecti, that contains conidia.

Rachis—Axis of an inflorescence, such as a barley or wheat head.

Rhizomorph—A cordlike strand of fungus hyphae.

Saprophyte—An organism that feeds on lifeless organic matter.

Scab—Crustlike disease lesion.

Sclerotium (pl. *sclerotia*)—Resting mass of fungus tissue, more or less spherical, normally having no spores in it.

Spore—A single to many-celled reproductive body in the fungi that can develop a new fungus colony.

Sporulation—Producing spores.

Systemic—A chemical that is absorbed and distributed throughout the plant.

Vector—An agent, insect, man, etc. that transmits disease.

Virus—Microscopic, transmissible, disease-causing agent (without cells); smaller than bacteria, viruses are composed of nucleic acid and protein.

Wettable powder (W.P.)—An inert powder that holds active material, usually insoluble in water but capable of forming a fairly stable suspension in water.

Wilt—Loss of freshness or drooping of plants because of inadequate water supply or excessive transpiration; a vascular disease interfering with water utilization.

Acknowledgments

A number of figures in *Section III, Major Insect and Mite Pests,* were adapted from the following sources:

PAGE 98, Figure 1, from James Richard Cate, Jr., *Ecology of* Erythroneura elegantula *Osborn (Homoptera: Cicadellidae) in Grape Agroecosystems in California* (PhD dissertation), University of California, Berkeley, no date.

PAGE 99, Figure 3, from J.F. Lamiman, *Control of the Grape Leafhopper in California,* California Agricultural Extension Service Circular 72, University of California, Berkeley, reprinted June 1937.

PAGE 101, Figures 4 and 5, Ibid.

PAGE 103, Figures 6 and 7, from Richard L. Doutt and John Nakata, "The *Rubus* Leafhopper and Its Egg Parasitoid: An Endemic Biotic System Useful in Grape-Pest Management," *Environmental Entomology,* June 1973.

PAGE 104, Figure 8, from James Richard Cate, Jr., *Ecology of* Erythroneura elegantula *Osborn.*

PAGE 113, Figures 1 and 2, and pages 114–115, Figure 3, from A. Earl Prichard and Edward W. Baker, *A Revision of the Spider Mite Family Tetranychidae* (San Francisco: Pacific Coast Entomological Society, 1955) and Edward W. Baker and A. Earl Pritchard, "A Guide to the Spider Mites of Cotton," *Hilgardia,* July 1953.

PAGE 152, graphs, from Dwight F. Barnes, *Notes on the Life History and Other Factors Affecting Control of the Grape Leaf Folder* (Leaflet E-616), U.S. Department of Agriculture, Agricultural Research Administration, April 1944.

PAGE 178, Table 1 and Figure 2, from Victoria Yoshiko Yokoyama, *Thrips Associated with Table Grapes and Their Effect on Fruit Quality* (PhD dissertation), University of California, Berkeley, no date.

On the preceding page:

Larvae of the western grapeleaf skeletonizer, *Harrisina brillians* Barnes & McDunnough, are pictured feeding on a grape leaf.

Section III—MAJOR INSECT AND MITE PESTS

Contents

Grape leafhopper adult.

The grape leafhopper, *Erythroneura elegantula* Osborn, is the most common pest of grapes north of the Tehachapi Mountains, infesting vineyards in the San Joaquin, Sacramento and Napa valleys, and appearing in the coastal valleys where it is not a common problem. Where it is prominent, every vineyard is infested, but many need not be treated because vines can tolerate fairly high populations without harm. A closely related species, the variegated grape leafhopper *Erythroneura variabilis* Beamer, also damages grapes, but it occurs only in southern California. (It is discussed in a separate section.) Both species are native to California.

In 1981 *E. variabilis* and *Dikrella cockerellii* (Gillette) were found infesting commercial vineyards in Fresno County. *D. cockerellii* reportedly is found only on wild grapes in southern California but it and *E. variabilis* infest vineyards in Arizona. The new infestations in Fresno County at this writing cause considerable concern.

The grape leafhopper has been reported in California since at least 1864. Before the development of organic insecticides in the 1940s, severe damage occurred in some years, followed by periods of low populations. Now, with more effective materials, losses are primarily the costs of treating plus the insect and spider mite disruptions that may occur as a result of treatment.

Actual pest damage varies according to location of the vineyard, variety, plant vigor, market use of the variety and season. While this insect is capable of defoliating vines by midsummer, such an event typically does not happen, even in untreated vineyards. On the average less than half of the wine or raisin grape vineyards require treatment, while most table grape vineyards require at least one treatment a year. Some vineyards never require treatment.

Description

The adult leafhopper is about 3.2 mm (1/8 inch) long, pale yellow with reddish and dark brown markings. The red and yellow patterns on the wings and head are more intense in diapausing adults than in summer and spring adults. In addition, the coloration on a diapausing female's abdomen is darker and covers a larger area (Figure 1).

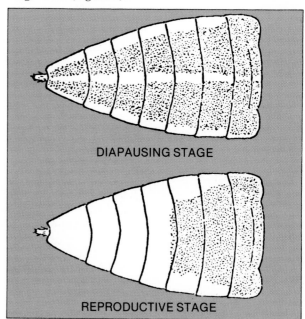

DIAPAUSING STAGE

REPRODUCTIVE STAGE

Figure 1. Seasonal abdominal patterns on females of *Erythroneura elegantula* in California.

ACTUAL SIZE: Grape Leafhopper			
-	-	-	-
	first	fifth	
EGG	NYMPHAL STAGES		ADULT

The grape leafhopper's eggs are laid singly in the epidermal tissue of both upper and lower leaf surfaces, although the lower is preferred. Each egg is minute, about 0.8 mm (3/100 inch) in length. The eggs can barely be seen with the naked eye on a smooth leaf like that of Thompson Seedless. In fact, even when numerous on a leaf, they are difficult to find with a hand lens. The presence of an egg is indicated by a small bean-shaped blister (Figure 2). The red eye of the developing nymph is visible shortly before hatch. Occasionally eggs turn dark or black and do not hatch; they are thought to be infertile or diseased. When eggs are parasitized they are reddish (see section on *Natural Control*).

When looking for leafhopper eggs, do not confuse them with the clear, tiny, round sap balls of various sizes found on the veins of leaves and young shoots in the spring.

The young leafhopper, called a nymph, emerges through a slit in the egg and leaf tissue. In this nymphal stage, the first of five stages, the leafhopper is almost transparent and colorless except for prominent red eyes. Each stage resembles the other except for increases in size and development of wings (Figure 3). The developing wings, in the form of wing pads, become more pronounced in the third, fourth and fifth stages. Nymph length ranges from about 0.8 mm (3/100 inch) when hatched to about 2.5 mm (1/10 inch) in the fifth stage. Although the nymphs molt five times, only the castoff skin of the fifth molt sticks to the leaf. Newly molted adults have white wings for two or three days until a characteristic pale yellow coloring and reddish and dark brown markings begin to develop. These new adults do not fly for a day or two.

Photo by Frank Skinner

Figure 2. Egg of the grape leafhopper in the tissue of a leaf. Actual size is about 0.8 mm (3/100 inch) long.

Nymph with developing wing pads.

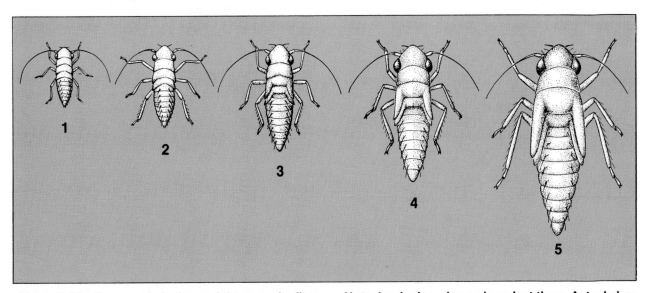

Figure 3. The five nymphal stages of the grape leafhopper. Note developing wing pads on last three. Actual size: Stage 1 is about 0.8 mm long (3/100 inch); stage 5 is 2.5 mm long (1/10 inch).

Injury

Both adults and nymphs feed on leaves by puncturing the leaf cells and sucking out the contents. Each feeding puncture leaves a white spot. Leaves that are fed on heavily lose their green color, dry up and fall off the vine. The injury does not interfere with the vine's function, but reduces photosynthetic activity by reducing effective leaf surface. Defoliation studies (in which leaves were removed manually) suggest that vines can withstand a 20 percent leaf loss a month after fruit set without adverse yield effect. Studies with varying levels of leafhopper have indicated a fairly high vine tolerance. Excessive fruit spotting in table grapes can lead to economic loss before any effect on soluble solids or yield is noted. During harvest high adult populations may annoy pickers by flying into their faces. Nymphs will also annoy by attempting to pierce the exposed skin of workers.

Seasonal Development

Grape leafhoppers overwinter as adults in a state of reproductive diapause, feeding on weeds in or near vineyards during warm weather. In cold weather they hibernate under leaves, dead grass, weeds and old paper trays, in brush and straw piles, in debris along ditches and fences and in alfalfa fields, dry tule ponds and old cotton fields. While these adults have reportedly been found up to a mile from the nearest vineyard, most are found in or close to a vineyard.

Grape leafhoppers remain in diapause until a photophase (day length) of about 11.6 hours is reached (about March 10); then the adults' reproductive organs begin to mature. Development of the eggs (ovigenesis) in females continues with feeding on plant hosts; the ovaries do not mature completely until the female has fed on grape foliage, so extended delay in egg development can occur if grape bud break is delayed. Adults may mate before feeding on grape, but the females will not lay eggs in other plants.

In March or April the adults move from their winter quarters to grapevines as soon as they can feed on green leaves, but they seek protection in trash or other cover under the vines in cold weather. Thus, they may feed on leaves during warm parts of the day and return to the ground in the evening. All adults do not move into the vineyard at once; they may take a month or so to migrate, gradually moving out of winter quarters.

The overwintered females lay eggs in grapes for up to six weeks, starting about mid-April. Eggs of the first brood hatch in 18 to 20 days (early May), depending on how warm the weather is. (Summer broods hatch in as little as ten days.) Development from the time of hatching to adulthood takes two to three weeks.

The nymphs feed on the leaf's underside. When disturbed, they often run crablike, sideways. Nymphs of the first brood are found primarily on the first six or eight basal leaves on a shoot where they hatched. First brood adults move midway on the cane to lay their eggs on fully developed new leaves, and the resulting second brood nymphs hatch in late June. The second brood overlaps into the third brood, with nymphs hatching until mid-September. The later broods are usually the most damaging because of their larger numbers. Adults maturing in late summer continue feeding on vines until frost. Diapausing populations begin developing about mid-August when the day length becomes less than 13.6 hours.

Brood development depends on temperature. An accumulation of approximately 980 day degrees above 10.3°C (50.5°F) is required to complete one generation. Studies in Fresno County, St. Helena in Napa County and at Berkeley show that an accumulation of 2,913, 2,276 and 1,477 day degrees governed the 3, 2½ and 1½ generations recorded in those locations.

Figures 4 and 5 diagram the seasonal development of grape leafhopper in the southern San Joaquin Valley, where three full generations usually occur.

Natural Control

Egg parasite of grape leafhopper. The most important natural enemy of the grape leafhopper in commercial vineyards is a tiny, almost microscopic wasp called *Anagrus epos* (Girault). Its progeny develop within the egg of the grape leafhopper, resulting in its death. Its size is about .3 mm (1/100 inch).

These parasitic wasps are particularly valuable because of their amazing ability to locate and attack grape leafhopper eggs. Also, their short life cycle permits them to increase far more rapidly than do leafhopper populations. Their nine to ten generations during grape growing season make them capable of parasitizing 90 to 95 percent of all leafhopper eggs deposited after July (see Figure 6).

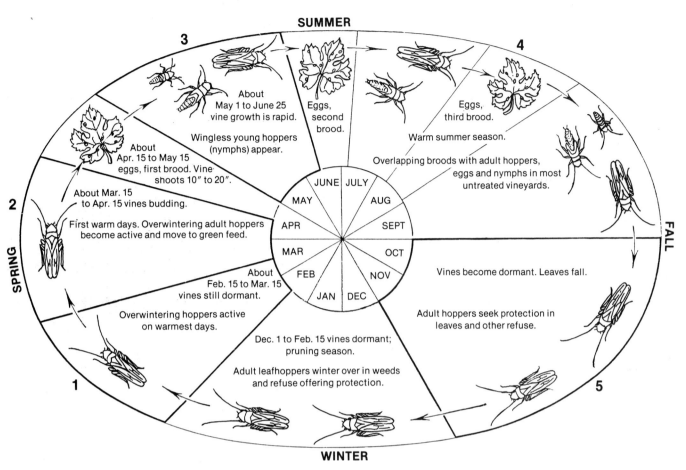

Figure 4. Seasonal life history of grape leafhopper.

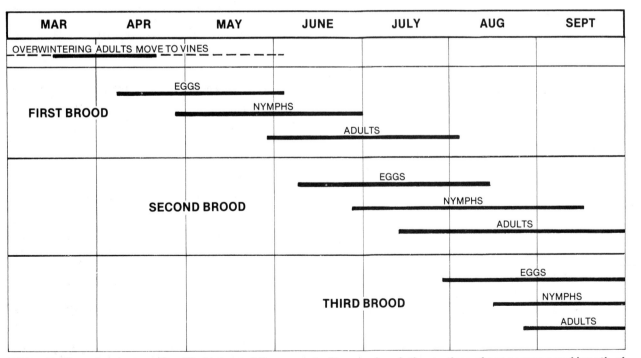

Figure 5. Diagram representing life history of the grape leafhopper in relation to time of appearance and length of various stages of the three broods. Note that as season advances overlapping of broods becomes more complex.

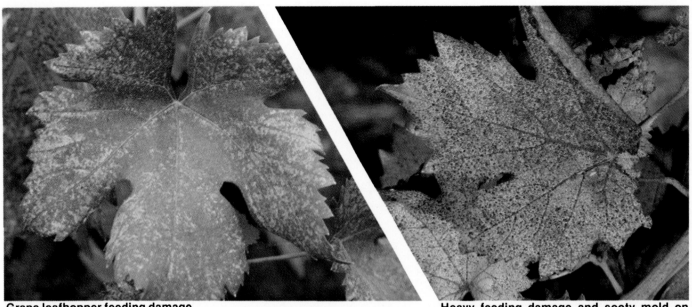

Grape leafhopper feeding damage.

Heavy feeding damage and sooty mold on leaf.

Leafhopper spotting on Calmeria table grapes.

Spotting on darker Ribier table grapes requires closer examination.

Leafhopper egg parasitized by *Anagrus*.

Parasite emerged from egg on right; unparasitized egg is left.

Anagrus adult female; terminal antennal segment clubbed.

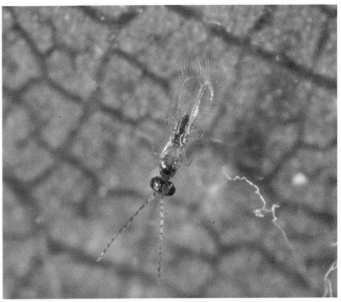

Anagrus adult male; terminal antennal segment not clubbed.

This parasite overwinters on wild blackberries, *Rubus* spp., on which it parasitizes the eggs of a non-economic, harmless leafhopper, *Dikrella* spp. These overwintering wasp populations tend to be along rivers that have an overstory of trees sheltering both wild grapes and wild blackberries. When the blackberries leaf out in February, the lush, new foliage apparently stimulates heavy oviposition by the *Dikrella* leafhoppers. The *Anagrus* parasites increase enormously on these eggs so that by late March and early April there is widespread dispersal of the newly produced *Anagrus* adult females. Fortunately, their dispersal occurs at the same time that grape leafhopper females begin to lay eggs (see Figure 7). Vineyards located within a five- to ten-mile range will usually benefit immediately from the immigrant parasites. Vineyards distant from actual refuges may not show *Anagrus* activity until midsummer or later.

Efforts to establish blackberry refuges near vineyards have been only partially successful. Studies in a

Tulare County vineyard showed that output of *Anagrus* parasites declined from an average of 2.5 parasitized eggs per leaf in 1966 to 0.8 in 1967 and 0 in 1968. This was attributed to a lack of a sheltering overstory of trees as occurs in natural systems. In full sun, plantings gradually became dry thickets, with only surface foliage that becomes tough and leathery.

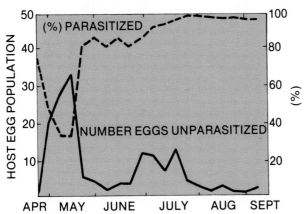

Figure 6. Normal seasonal pattern of grape leafhopper egg parasitization by *Anagrus* in a vineyard close to wild blackberry thicket. Parasitism began early with development of first brood and reached a high percentage during second and third broods.

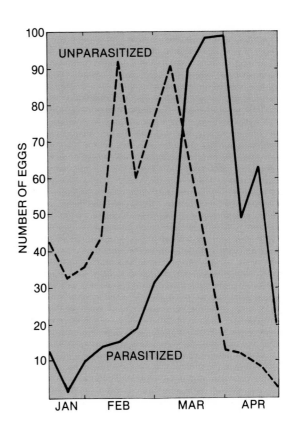

Figure 7. Annual production of *Anagrus* from *Dikrella* eggs. Note that the peak emergence of adult *Anagrus* occurs at time grape leafhoppers begin laying eggs in vineyards.

Dikrella is a shade- and moisture-loving insect that breeds on leaves inside berry vines where it is generally cooler and humid. Old maturing thickets, therefore, become less desirable *Dikrella* habitats and are, consequently, poor producers of *Anagrus*. A vineyard located in a natural dispersal area will, therefore, receive early, important activity from these parasites. Growers with vineyards too distant for early benefit should follow the guidelines of nymphal counts for treatment (see section on *Management Guidelines*).

Parasite of grape leafhopper nymphs and adults. Another parasite, *Aphelopus comesi* (Fenton), a dryinid wasp, attacks third, fourth and possibly fifth instar leafhopper nymphs. *Aphelopus comesi* has a relatively long life cycle well attuned to the grape leafhopper's life cycle. In one case in the Fresno area 70 percent parasitism was recorded. Generally, however, parasitism by *A. comesi* rarely exceeds 10 percent. This parasite places its tiny egg between the nymph's second and third abdominal segments where it remains undeveloped until the leafhopper nymph molts to the adult stage.

The first larval instar of the developing parasite is recognizable as a poorly defined, light colored, oval mass under the dorsum of the adult leafhopper's first abdominal segment, usually to the left side of the body (Figure 8A). During the parasite's larval development, the host appears normal, except for an elongating larval sack (the thylacium) that gradually protrudes from the abdomen with each molt (Figure 8B, C, D). By its fifth instar the parasite larva has well developed mandibles and removes the reproductive organs of the adult leafhopper so it cannot reproduce. The parasite larva then drops to the soil, spins a cocoon in which it remains for 14 days, pupates, and emerges as a sexually active adult in five days after it pupates.

Adult female and male parasites differ in color and size. The female is about 1.8 mm (7/100 inch) long, is brown to light tan and has a small stinger projecting downward from the tip of the abdomen. The male, about 1.5 mm (6/100 inch) long, is black with white mandibles, face and legs.

General predators of grape leafhopper nymphs and adults. Several general predators prey on leafhopper adults and nymphs of all stages all season. Among the most abundant of the general insect predators are *Chrysopa* sp. (green lacewing), *Orius* sp. (minute pirate bug), *Nabis* sp., *Hippodamia convergens* (lady

bird beetle), *Geocoris pallens* (big eyed bug), and *Hemerobius* sp. Of the different species of spiders that prey upon grape leafhoppers, the most abundant are Thomisidae and Araneidae.

First instar nymphs also are attacked by the predaceous mite *Anystis agilis* (Banks), particularly in the Napa Valley. *Anystis* is a general predator that feeds on spiders, mites and insects. It is relatively indiscriminate about its prey (it cannibalizes its own nymphs and eggs) as long as the prey is relatively soft. Counts in St. Helena vineyards suggest that two generations occur, one in late spring and one in early fall. *Anystis* is a large, red, long-legged, rapidly moving mite, commonly called the whirligig. Eggs are large, round, yellow-brown and finely granulated. About six to twelve eggs are grouped on the ground or grape foliage and appear enclosed in a brownish-yellow mass. *Anystis agilis* reportedly spins silken cases or "cocoons" in which to rest between molts.

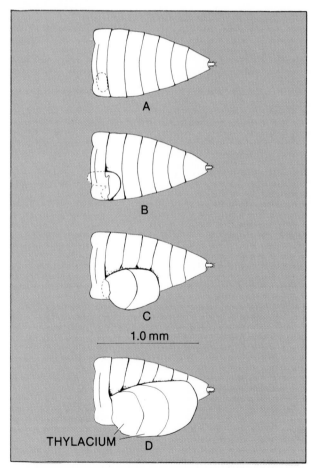

Figure 8. Developmental stages of larvae of *Aphelopus comesi* on dorsal abdomen of grape leafhopper adult. Parasite larva develops in sack called thylacium.

Reports of diseases as mortality agents of grape leafhopper are sketchy at best. It has been suggested that its eggs are diseased when they turn black and do not hatch. As there are times when black eggs are numerous, this phenomenon needs study. There are also unconfirmed reports of grape leafhopper population declines caused by a fungus, *Beauveria bassiana*; this too needs study.

Monitoring Guidelines

Leafhoppers prefer vigorous vines. Although leafhoppers naturally tend to disperse, the heaviest populations are normally found on end vines and on outside rows. This is partly because these vines are usually the most vigorous and therefore the most attractive. It also is partly because of the border or boundary effect. Fortunately, the most vigorous vines can tolerate the highest populations.

Weak vines may show only half the leafhopper populations of stronger vines. This contrast is more evident in summer broods than during first brood because the weakest vines stop growing about mid-June, and leafhoppers prefer to lay eggs in newly matured leaves. Thus, once vine growth ceases, egg laying declines. With vigorous vines that continue growing into late summer or fall females have a constant supply of new leaves in which to lay eggs.

Surveying for leafhopper nymphal populations. Leafhopper populations in a vineyard of standard spacing (vines 8 feet apart in rows on 12-foot centers and approximately 640 feet long) may be surveyed as follows:

(1) *Choosing the vine.* At each end of the block pick every fifth row, go in to the fifth to tenth vine and select *one* leaf for examination (avoid outside rows and the two end vines, as these usually show higher-than-normal populations).

(2) *Locating and selecting the leaf for examination.* First brood: select one of the basal leaves on the shoot (second through sixth leaves give the highest counts). Second brood: select a midshoot leaf.

Take leaves from the east side of rows planted north and south or from the north side in vineyards planted east and west. The shadier side usually has the heavier count. Look for evidence of feeding damage. Before counting on any particular day, examine a few leaves to determine which are likely to give the highest counts. Then, once leaves are selected and the count begun, don't discard any leaf, even if no nymphs are found on it.

(3) *Averaging counts.* For any block count nymphs on a minimum of 15 leaves. Add the nymph numbers and divide by the number of leaves examined to obtain the average number of nymphs per leaf. Keep separate figures for each block.

(4) *How often to count.* Two or three counts for the first brood and additional counts every seven to ten days beginning with the second brood are sufficient. Should populations appear to be changing rapidly, additional counts may be necessary. Remember that early nymph populations are small, so expect their numbers to climb with successive counts. These young nymphs do little damage so a few days' delay in counting or treating does little harm.

This system for surveying nymphal populations was developed for vineyards of 10 to 40 acres and takes only two to three minutes per acre. Larger commercial acreages require a slightly different system. Because it is faster to drive between sample locations instead of walking, three leaves are examined at every thirtieth row at the end of each block (10 to 15 vines in) rather than one leaf every fifth row. With this system only one leaf per acre is examined. One person can survey 500 to 1,000 acres per day, although the number of nymphs per leaf will influence the total time needed for counting.

Management Guidelines

Cultural effects. While the susceptibility of vineyards to the Pacific spider mite may be alleviated by cultural practices to increase vine vigor, leafhopper populations cannot be managed by these means. The maximum productive crop capacity is obtained from moderately vigorous vines and these are attractive to leafhoppers. Overly vigorous vines are not desirable either for their productive potential or their susceptibility to leafhoppers. Fortunately, vines growing with moderate vigor can tolerate fairly high populations without adverse effects.

Certain varieties are likely to suffer higher leafhopper populations than others. The table grape varieties Emperor and more particularly Ribier, in contrast to Thompson Seedless, are likely to become infested with high populations. Similar observations have been made on various wine varieties. Moreover, late producing varieties are more likely to favor leaf-

Blackberry leafhopper egg is typically laid in leaf vein (note arrow).

Blackberry leafhopper *Dikrella* adult.

Blackberry leafhopper nymph. Note dark hairs.

Adult grape leafhopper parasitized by *Aphelopus comesi* (arrow).

hoppers than earlier maturing varieties. The former produce a continuation of newly matured leaves that are favored by leafhoppers for depositing eggs. Mass movement out of earlier maturing varieties into nearby late growing varieties is commonly observed. These migrations may result in the need for treatment of the late variety, particularly the rows bordering the earlier variety. This kind of movement is even noticeable in the same vineyard where hoppers are attracted to the more vigorously growing vines near irrigation pots or at the end of irrigation runs.

If all weeds and trash in the vineyard are disced in before spring, the adults lose their protection and feeding sites. In general farming areas such a practice has less value as the adults will simply move to an adjacent alfalfa field, grain crop, etc., or weedy area.

"Trap crops," like small-seeded horse beans and barley, have been grown in strips two to three feet wide in every tenth vineyard row to attract overwintering adults before vines leaf out. The trap crop is then treated with an insecticide to control the adults. Little actual benefit has been seen, however, especially since overwintering adults are difficult to control with the chemicals available. As with winter cleanup, benefit is likely to be appreciable only when extensive areas are handled.

Leaf damage and vine tolerance. Assessing the leafhopper population or its damage is not simple. Helpful approaches include:
- Rating nymphal counts while assessing potential of flying adults.
- Rating leaf damage and loss.

Anystis agilis attacking grape leafhopper nymph.

Nymphal counts are easy and rapid. Adult populations can only be described in such relative terms as low, moderate or heavy, because most fly from the leaves selected for nymphal counts. Nymphal counts alone are adequate for deciding first brood treatments because this brood's peak lasts a short period. Nymphal counts for second or third broods are not as good a guide for control decision because these summer generations last two months or longer with less predictable populations. Furthermore, a brief peak of 30 nymphs per leaf in the first brood may do much less damage to a vine than a population of 15 per leaf extending through summer.

A fairly accurate way to estimate the impact of a nymphal population on the vine is to multiply the number of nymphs per leaf by the number of days of exposure. The resulting figure is called "nymphal days." Table 1 gives an example of how to accumulate nymphal days from the beginning of the second brood. The impact of adult populations is not included; it must be dealt with by observation.

Another way to assess population is to observe leaf damage and loss. Trials with manual leaf removal have shown that vines can lose up to 20 percent of their leaves without any yield or maturity loss, provided the leaves are not removed until about a month after fruit set. Inasmuch as fruit set normally occurs about June 1, vines can tolerate a 20 percent leaf loss after July 1. Second brood nymphs generally do not begin hatching until late June or early July and this brood will not damage leaves until mid-July, when damage can exceed 20 percent.

TABLE 1. Weekly Nymphal Leaf Counts and Days and Their Accumulation.*

Date	Nymphs Per Leaf	Nymphal Days	Accumulated Nymphal Days
First brood:			
May 28	6.8		
June 4	7.8		
June 11	7.0		
June 18	4.6		
June 25	2.4		
Second and third broods:			
July 2	1.7		
July 9	5.2	24†	24**
July 17	18.0	93	117
July 23	14.0	96	213
August 4	5.9	119	332
August 11	0.7	23	355
August 19	1.7	10	365
August 25	2.1	11	376

*Data from Thompson Seedless vineyard.

†This figure is obtained by averaging the July 2 count of 1.7 nymphs per leaf and the July 9 count of 5.2 nymphs per leaf, and then multiplying by the number of days between July 2 and July 9.

$$\frac{1.7 + 5.2}{2} = 3.45 \text{ nymphs per leaf}$$

July 2 to July 9 = 7 days
3.45 nymphs per leaf × 7 days = 24 nymphal days

**Nymphal days calculated for each period is added to the previous figure to give the accumulated nymphal days.

Both experiments and observation of normal practices indicate that vines can tolerate some leaf loss without damage to yield. Growers often remove a substantial number of leaves by discing or by mowing the tips of the shoots before discing. Table grape growers remove leaves around the clusters on red and black varieties to enhance coloring and may cut canes on one side of the trellis, thereby also reducing functional leaf surface.

Effect of *Anagrus* on appraising leafhopper population. The presence of the parasitic wasp *Anagrus* is easily recognizable in the vineyard. Parasitized leafhopper eggs turn a brick red before the adult wasp emerges. These eggs are clearly visible, especially on such smooth-leafed varieties as Thompson Seedless. Occasionally, if there are few parasitized eggs, the parasite's presence must be confirmed with a hand lens.

If parasitized eggs are found in the first brood, the possibility of the leafhopper population remaining within tolerance levels is good, especially with Thompson Seedless. Even if the population is above the tolerance level, a decision about treating the first brood can be delayed to give the parasite an opportunity for control.

If *Anagrus* does not reach the vineyard until the summer broods, it cannot parasitize a high percentage of leafhopper eggs, although its long term activity is still beneficial.

Even when *Anagrus* is absent, leafhopper populations may frequently remain well below tolerance levels because of such other control factors as early vine maturity. However, presence of *Anagrus*, especially during the first brood, should always be regarded as a significant natural mortality factor (although chemical control measures may still be warranted).

Treating broods. Overwintering adults: In the past, growers were sometimes encouraged to treat vines to kill the overwintering adults before April 15 to prevent egg laying for the first brood. This is no longer suggested except where overwintering adults themselves are so damaging as to require treatment. In the usual vineyard situation low adult numbers usually mean that the first brood will be light, but high numbers do not necessarily mean a heavy first brood. Also, because adult movement from overwintering quarters to vineyards proceeds for an extended time and the adults are difficult to control, treatment before egg laying provides only partial population reduction.

First brood nymphs: Normally first brood nymphs affect a small portion of total leaf surface on a vine and should only be controlled if they become so numerous as to cause heavy damage. These nymphs remain on the basal six to eight leaves where they hatched. New leaves remain free of leafhoppers until the nymphs mature and move to them. Feeding by the nymphs will not stunt a vine, but with high numbers (20 to 30 per leaf), functional leaf surface may be reduced by late May or early June.

If *Anagrus* is active in the vineyard (red eggs present) at this time, avoiding first brood treatment may allow the parasite to increase to levels that will make it even more effective during the critical second brood.

Second and third broods: These broods can be the most damaging, excessively spotting table grapes and reducing wine or raisin grape yields. Under the most severe conditions they may defoliate the basal and midcane leaves by mid- to late July, although such destruction is rare.

Once second brood nymphs begin hatching, the vines will support all stages of leafhoppers. In general no one time is preferable for treatment. If treatment can be delayed into the second or third brood, the parasites will achieve maximum usefulness, and, if treatment is necessary, one treatment will often suffice.

Tolerance level of Thompson Seedless. In trials with Thompson Seedless in the San Joaquin Valley reductions in yield or sugar did not occur even when first and second brood population levels respectively reached peaks of 20 and 25 nymphs per leaf. The levels attained were within the range of population that can be tolerated by the Thompson Seedless vines without adverse effects.

Nymphal day counts can also be used to estimate population damage potential. Calculations should begin with the appearance of the second brood, usually in early July. At this time vines apparently tolerate exposure of about 600 nymphal days without affecting yield or sugar content. Considering nymphal counts without regard to the length of time the vines are subjected to feeding is not as reliable as using the nymphal day summation. Neither method, however, surveys the adults, which must be appraised by the amount of leaf damage and loss. Table 2 summarizes tolerance levels on Thompson Seedless for wine or raisins.

Although tolerance studies have been limited to Thompson Seedless, other varieties of similar vigor would probably show similar tolerance levels. In general strong, vigorous varieties will have greater tolerance than weaker ones.

Table grape tolerance levels. Excessive spotting or leafhopper droppings determines the economic damage level on table grapes (and this occurs long before any reduction in sugar or yield can be measured). As adults and nymphs feed on the undersides of leaves their excretions simply fall on the tops of the berry clusters.

Such spotting usually starts with the second brood's hatch. Potential damage depends on the length of exposure which varies with the harvest date. Early maturing varieties, such as Perlette and Cardinal, are exposed only a short time and so can tolerate fairly high leafhopper populations. Late varieties, such as Emperor and Calmeria, accumulate spotting for more than three months, which means they can tolerate lower populations. Damage to these late varieties, however, may be lessened by any fall rains that wash off spotting.

Grapes with a deep red color obscure spotting and have higher population tolerances. Spots cannot be seen at all on fully ripened Ribier, a black grape, although the berries may be so coated with leafhopper droppings as to feel rough.

There are no definite standards for "excessive spotting." The U.S. Standards for Table Grapes states that the berries are damaged "when the appearance is materially affected by the presence of leafhopper residue." Because this description leaves the determination of damage to personal opinion, arbitrary maximums have been set at 75 spots per square centimeter for white table grapes and 100 spots per square centimeter for red or black table grapes. Table grapes have been shipped with spotting levels above those without adverse judgment from inspectors. To monitor spotting take a berry from the most exposed area of shoulders of at least 10 (preferably 20) bunches. Use a hand lens to count the spots in an area approximately 6.4 mm (1/4 inch) in diameter on each berry. Determine the average number of spots on all berries checked. Multiply this figure by 3.16 to convert to square centimeters.

TABLE 2. Guidelines for Leafhopper Tolerance Levels in Thompson Seedless for Raisins or Wine and in Table Grapes.

	First Brood	Second and Third Broods
Thompson Seedless for raisins or wine:	20 nymphs per leaf	10 to 15 nymphs per leaf Note: Do not permit more than 20% leaf loss. or 600 nymphal days from beginning of second brood until harvest.
Table grapes*:	10 to 15 nymphs per leaf	Early varieties: 10 nymphs per leaf Midseason varieties: 5 to 10 nymphs per leaf Late varieties: 5 to 8 nymphs per leaf or 300 nymphal days from beginning of the second brood until harvest.

*Tolerance levels are influenced by the presence of *Anagrus.* If the latter is very active (numerous red eggs on leaves), the numbers given (nymphs per leaf) are conservative and treatment decisions should be based more on amount of fruit spotting.

Little spotting takes place with the first brood unless nymphal populations are more than 15 per leaf. Therefore, spotting counts normally start with the hatch of second brood nymphs—usually in early July. As spotting accumulates, some weathers off and some is diluted out by the growth of the berries. Thus, with a light population of three nymphs per leaf or below, spotting may remain static, weathering off about as rapidly as deposited. Heavy populations may lead to rapid accumulation of spotting up to or past the tolerance level. But when effective treatment or effective parasitism reduces leafhoppers, weathering will reduce spotting considerably.

Berry spotting counts and leafhopper populations are obviously related, but there are no simple methods to correlate them. Adults produce more spotting than nymphs, but adults cannot be counted. Usually the nymphal population reflects the incipient adult population, but instances do occur where adults live much longer than normally and produce spotting not accounted for in nymphal exposure. Additionally, the relationship of nymphs to spotting is not proportional; that is, 10 nymphs produce more than twice as much spotting as five. The same level of exposure to nymphs in July and early August will produce more spotting than in late August and September.

The best index of leafhopper population to berry spotting is nymphal day accumulation. About 250 nymphal days are regarded as the tolerance for table grapes; in various trials this figure has ranged from 200 to 400. Table 3 gives an example of berry spotting counts and accumulated nymphal days in an Emperor vineyard. Note the dropoff in spotting with population decrease. This vineyard was not treated.

Those who do not wish to keep a nymphal day accumulation should make periodic nymphal counts and general observations to gain a practical idea of leafhopper impact. Tolerance levels for table grapes are shown in Table 2.

Finally, note that the guidelines given in Table 2 are based on averages. Watch for wide variations. For example, a vineyard average may be well below the tolerance level, but parts of the vineyard may have high and damaging population levels that may call for treatment decisions. This is particularly important when assessing table grapes where tolerance levels are more restrictive.

Date	Nymphs Per Leaf	Nymphal Days	Accumulated Nymphal Days	Berry Spotting Spots per cm²
TABLE 3. Nymphal Counts, Nymphal Days and Berry Spotting.*				
First brood:				
June 14	1.1			
June 21	1.2			
Second and third broods:				
July 6	.65			
July 19	7.0	50	50	
July 31	8.5	93	143	78
August 15	3.1	87	230	134
August 28	3.5	43	273	99
September 14	.05	30	303	87
September 28	.30	2	305	29
October 11	.00	2	307	37

*Data from Emperor table grape vineyard.

Pacific mite outbreak spot in Thompson Seedless vineyard.

Pacific mite female.

Willamette mite female.

Willamette mite egg showing hairlike papilla (arrow).

ACTUAL SIZE: Spider Mites
·
PACIFIC
·
WILLAMETTE

Spider mites have been serious pests of Central Valley grapes for several decades. Reasons for this development have not been fully documented, but it is believed that the use of synthetic organic insecticides has upset natural controls. Serious spider mite problems in vineyards can be reduced, however, by judicious use of pesticides.

Two spider mite species are commonly abundant on grapes in California: Pacific spider mite, *Tetranychus pacificus* McGregor, which deserves serious attention, and Willamette spider mite, *Eotetranychus willamettei* (Ewing), whose populations only occasionally become large enough to cause concern. The twospotted spider mite, *T. urticae* Koch, is only occasionally found on grapes.

Information here about their distribution and abundance has come primarily from Fresno, Kern and Tulare counties in the southern San Joaquin Valley, although problems are increasing in the Napa Valley. Spider mites are not considered grape pests in southern California, at least in the desert areas. An outbreak of Willamette mite in 1979 in central coastal valley areas caused considerable concern.

Distribution and abundance of Pacific and Willamette mites are shifting because of pesticide use and large scale planting of wine varieties. In general, however, environmental conditions and soil types strongly influence mite populations. For example, in western Fresno and Tulare counties (Caruthers and Dinuba areas) soils are generally alkaline and light in texture, contributing to the hot, dry, dusty vineyard conditions that favor the Pacific mite. In eastern Fresno and Tulare counties, where heavier soils prevail, vineyards are more humid and less dusty, and the Pacific mite is considerably less favored. It is also probable that spider mite nutrition (which affects reproductive rate and longevity) may depend upon how soil type affects grapevine chemistry.

Use of insecticides in the hot, dry areas of western Fresno and Tulare counties has accentuated problems with the Pacific mite, while in the eastern areas it is rarely a pest in spite of heavy insecticide treatments for leafhoppers. Generally, Willamette mite populations are predominant in the eastern areas, and they, too, are favored by dusty vineyard conditions.

Vineyards in which grass culture is practiced tend to have fewer problems because these vineyards are slightly cooler, more humid and less dusty. Grass culture usually improves water infiltration so that vines are better supplied with moisture. Improving water intake is important because moisture stress apparently favors spider mites.

Description

Egg stage. Both the Pacific and the Willamette mite deposit eggs singly on the undersurface of a leaf, particularly along the midrib and veins. Eggs are found on the upper leaf surface only when population densities are high. Pacific mite also lays eggs in its webbing. The egg in both species is spherical, but that of Pacific is slightly larger. When first laid, the egg of each species is translucent; that of Pacific is slightly amber. With incubation the egg becomes opaque; the embryo's reddish eyespots become distinctly visible shortly before the larva hatches. The Willamette mite's egg has a fine papilla (hair) that tapers at the top; no papilla is found on the Pacific or twospotted spider mite egg.

Larval stage. The newly hatched larva has six legs. "Food spots" form soon after feeding—on the dorsum (back) of the Pacific mite and on each side of Willamette.

Protonymph stage. In this eight-legged stage food spots become more apparent in Pacific mite, but are inconspicuous in Willamette.

Deutonymph stage. The mite is similar in appearance to an adult female but smaller. Males of both species are distinguishable, but the distinction is not as clear cut as in the adult stage. Food spots become conspicuous in Willamette mite.

Adult stage. Upon emergence, adult Pacific mite females are 0.5 mm (2/100 inch) long and are almost devoid of food spots, but as feeding begins the spots become more distinct although the pattern varies. Usually there are two large, diffused spots forward and two smaller spots on the rear. This pattern is less evident early in the season. The rear spots are diagnostic for field identification to dispel confusion with twospotted spider mite (see Figure I). Adult Pacific spider mite females vary in color from slightly amber or greenish to reddish; later in the season and at high densities they become reddish, a coloration not to be confused with the deep orange of resting state (diapause) females.

Adult Willamette mite females are 0.5 mm (2/100 inch) long and are usually pale yellow with small black dots along the sides of the body behind the eyes (sometimes as many as four distinctly visible spots occur on each side).

Adult males of both species are easily recognized by their pointed abdomens and smaller size (about one-half of the size of the mature female). The males aggressively attend deutonymph females, chasing off other males. A hand lens and practice are necessary to distinguish the two species. Mixed populations of both species may occur.

Sometimes microscopic examination is desirable. Growers may request their farm advisors or pest control advisers to collect specimens (males included) in vials with 70 percent alcohol for study by the appropriate university or State Department of Food and Agriculture taxonomist.

For pest control advisers or farming operations that possess the proper optical equipment for making their own identifications, here is the procedure:

Adult females are taken from the leaf and placed in a small drop of Hoyer's mounting medium (recipe follows) on a slide, dorsal surface up. A cover slip is applied and pressed down lightly. To speed clearing

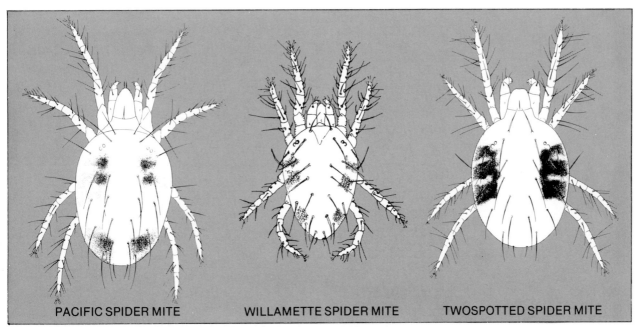

Figure 1. Spotting pattern of Pacific, Willamette and twospotted spider mite females.

of the specimen, the slide can be heated *gently* on a desk lamp or hot plate.

With a microscope on 45X magnification (objective power) Willamette mite is distinguished from Pacific by its possession of an extra pair of setae, the clunals, near the posterior end of the adult female (see Figure 2). To determine the presence or absence of the clunals on a specimen the terminology and the setal pattern in Figure 2 should be studied carefully.

Species identification can also be made by mounting a single male, but it must be manipulated onto its side to examine its aedeagus or copulatory organ. This is accomplished by carefully moving the cover slip so that the specimen rolls on its side. The cover slip is then pressed down slightly to make the aedeagus protrude. (Lining up the small red eyes before pressing down facilitates proper lateral orientation.) It is best to remove males directly from the leaf for mounting, because those that have been in alcohol

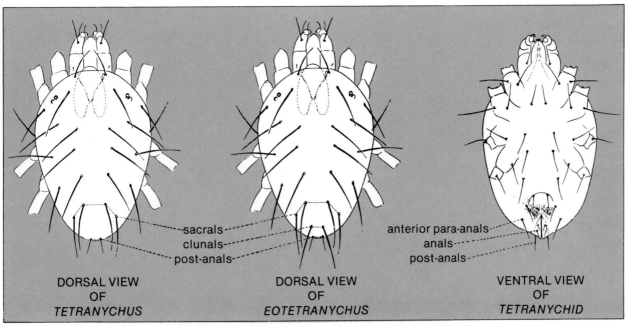

Figure 2. Dorsal and ventral setae pattern of *Tetranychus* and *Eotetranychus* spider mites.

are more difficult to position; eye color is also lost in alcohol-stored specimens.

Using the descriptions provided in Figure 3 and with proper mounting and examination of the aedeagus, Willamette, Pacific and twospotted spider mite males are easily compared and distinguished.

Recipe for Hoyer's Mounting Medium:
Mix the following ingredients in sequence:

Distilled water	50 ml
Gum arabic (amorphic)	30 grams
Chloral hydrate	200 grams
Glycerine	20 ml

Solid ingredients should be completely dissolved before adding succeeding reagents. This is aided by continuous stirring with a magnetic stirrer. The gum arabic is the most difficult ingredient to dissolve and may take up to a week of continuous stirring to get a complete mixture. Crystalline gum arabic is much easier to dissolve than is powdered gum arabic. The final product should be filtered through clean cheesecloth to remove bits of wood or other detritis from the gum arabic.

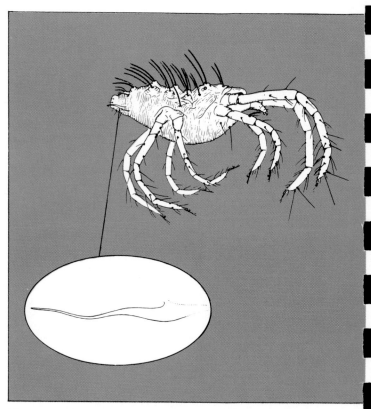

Figure 3. Male spider mite and its reproductive structures.

Typical Pacific mite leaf burn on top of grape vine.

Closeup of Pacific mite webbing.

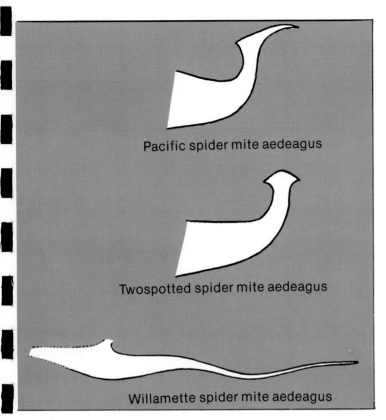

Pacific spider mite aedeagus

Twospotted spider mite aedeagus

Willamette spider mite aedeagus

Injury

Feeding by small colonies of both species produces small yellow spots on upper leaf surfaces. Yellowing of an entire leaf is characteristic of high densities of Willamette mites, while a sickly bronze discoloration is produced by high densities of Pacific mites. Later, especially in hot weather, the leaves turn dry and brown. Distortion of shoot tips is also characteristic of Pacific mite injury.

Depending upon the degree and timing of the injury, fruit quality or maturity may be adversely affected by Pacific mites because of loss of productive leaf surface. Also, the following year vine growth and crop are sometimes affected. Shedding of leaves occurs with heavy Pacific mite populations. Rarely, heavy infestations of Pacific mites have killed grapevines.

Willamette mites seldom produce economic damage to grapevines in the San Joaquin Valley and only infrequently require control. They may cause berry ambering of such table grapes as Thompson Seedless, because of exposure of the bunches to sunlight. Vigorous vineyards seldom require treatment. In

Typical Willamette mite yellowing damage is seen on vine, left, and in closeup, right.

fact, it may be important to preserve Willamette mite as an alternate food source for predaceous mites (discussion follows).

On rare occasions large numbers of overwintering Willamette mites may cause heavy leaf damage on young shoots in spring, but this is usually spotty and the vines outgrow the injury without noticeable economic losses. On the other hand, large populations of overwintering Pacific mites at bud break can cause more serious injury.

Seasonal Development

Pacific and Willamette mites overwinter under grapevine bark as mature females. This resting state (diapause) is induced by a combination of short day lengths and lower temperatures during early fall and probably is enhanced by foliage maturity. Observations indicate that Willamette mite requires shorter day length to induce diapause as it still actively reproduces later in the fall than does Pacific mite.

Diapause is also induced in Pacific mites by the foliage injury that their own high populations produce, and it is not uncommon to find their mature females in diapause under the bark on injured vines during summer. Whether these early diapausing females remain under the bark and overwinter is unknown. Neither is it known what effect feeding injury by either Pacific or Willamette mites has on diapause induction of the latter. Female Pacific mites in diapause exhibit a deep glistening orange coloration and lack of food spots. These females differ from those that become reddish as foliage quality declines. Diapausing Willamette mite females are a glistening lemon yellow and also lack food spots.

Overwintering females of both species move on to young foliage when grape buds break in spring, with Willamette mite apparently more active during cooler temperatures. Both species produce many generations each year. During favorable periods population explosions result because the females can daily lay more than eight eggs, about two-thirds of which develop into egg-laying females in ten days or less.

Natural Control

Predaceous mites on grapes. The most important predatory mite found in California vineyards is *Metaseiulus occidentalis* (Nesbitt), which under certain conditions (abundant prey and favorable temperatures) has enormous capacity for population increase. Another species, *M. mcgregori* (Chant), has reportedly attacked Willamette spider mite in commercial vineyards in the Lodi area, but its effectiveness is unknown. Other predatory mite species are only occasionally found on grape foliage and exert little, if any, control of spider mites.

Capacity for increase is not the only criteria for determining whether a given predator can effectively control a pest. Other important factors are: predator-prey distribution patterns in the vineyard, availability of alternate prey or food for the predator and viticulture practices that affect spider mite population levels.

Description. Predaceous mites are easily distinguished in the vineyard with practiced use of a good hand lens. The pregnant, egg-carrying female is pear-shaped and slightly larger than an adult female spider mite. Color varies, depending upon the recency of feeding. Predaceous mites are often slightly reddish if Pacific mites are the prey and slightly yellowish if Willamette mites are.

Eggs of predaceous mites are football-shaped compared with the spherical eggs of spider mites. Freshly laid, the eggs are clear and then gradually take on a milky appearance, becoming more opaque before hatching.

There are four postemergent stages; the first is a six-legged larva and the remaining three are eight-legged. In order of development they are called the larva, protonymph, deutonymph and adult stages.

Development. The female is usually found along leaf veins or wedged in vein angles where she prefers to lay her eggs in a loose group. Eggs are also laid in the webbing of Pacific mite colonies. After hatching, the larva wanders awkwardly in search of food and after feeding it becomes more active. The larval stage attacks all stages of prey, but prefers spider mite eggs. Before molting to the protonymph stage, it passes a short resting period.

The protonymph moves quickly while searching for prey, successfully attacking all stages of prey, although it may not eat the entire adult. More than one predator is sometimes seen feeding on the same prey. The protonymph stops moving for a short time before molting to the deutonymph stage. The deutonymph is similar to the protonymph in habits and activity.

Unlike the protonymph and deutonymph, the adult female, if not hungry, is less active, spending considerable time in protected angles of leaf veins. The male adult, which resembles the nymphal stages, often attends the female deutonymph while she is quiescent and ready to molt. Mating usually occurs after the adult female emerges. With prey available and favorable conditions, *M. occidentalis* is capable of developing from egg to egg-laying female in five days.

Metaseiulus occidentalis overwinters primarily under the buds of grapevines as a mated, adult female. As with spider mites, diapause for overwintering survival is induced by short day lengths and cool temperatures in fall. It is important that prey be available for these predators in fall. Without food reproduction stops and the predator population ceases to produce the proper stage for diapause induction that occurs during the immature stages.

The resulting female mates, but lays no eggs and only occasionally feeds. She remains inactive until cooler weather induces migration to the overwintering sites under bud scales. There she remains until the following spring. The male does not overwinter. In the spring the emerging female searches the expanding foliage for prey and begins laying eggs, the number depending upon availability of prey. When a well distributed population of predators successfully overwinters in a vineyard, spider mites are adequately controlled unless subsequent pesticide treatments disturb predator-prey interactions.

Alternate prey. Besides spider mites, *M. occidentalis* preys upon other mites, such as eriophyids, tydeids and possibly tarsonemids. These other mites help to support and maintain predator populations, especially in the fall when spider mites are scarce. Tydeid mites are particularly important in this respect; without sufficient numbers of them in the fall only low populations of predatory mites could overwinter. Thus, abundant tydeid mites indirectly influence effective predation of Pacific mites.

Tydeid mites are not pests of grapes; they feed primarily on pollen and to some extent on fungi. Their secondary feeding on leaf tissue produces no discernible injury. They are difficult to see on the leaf without a microscope; the stages (adults, immatures and eggs) are about one-third the size of equivalent spider mite stages. Characteristically these tiny mites move as fast backwards as forward. Their quiescent (molting) stages are found by searching along leaf veins and between vein angles. Their small, oval-shaped and stalked eggs are usually laid on leaf hairs. They eat their own eggs.

Two species of tydeids have been reported in commercial vineyards in the San Joaquin Valley: *Pronematus anconai* (Baker) and *P. ubiquitus* (McGregor). Little is known about their abundance and distribution on grapes in California. A different tydeid, *Triophtydeus* sp., has been found on grapes in the north coast area; little is known about its role as a alternate prey for *M. occidentalis*.

Tydeid mite populations, in contrast to Pacific mites, persist on foliage late into the fall. In Thompson Seedless vineyards in Fresno County's Caruthers area they were observed to constitute the main source of food for predators in fall because Willamette mites were scarce. Tydeid mites are found overwintering under bud scales, presumably in the diapause state. *Metaseiulus occidentalis* occasionally has been observed to prey upon tydeids under buds during winter.

Vineyards in eastern Fresno and Tulare counties appear to present more favorable conditions for tydeid mites than do vineyards in the western parts, particularly if there are weedy conditions.

It has been observed that tydeid mites are rare in Caruthers area vineyards where pesticide applications cause Pacific mite populations to flare up. It is not known why this occurs. The unavailability of tydeids reduces fall predator populations. Vineyards in the same area in which Pacific mites have not been disrupted by pesticides contained sufficient numbers of tydeid prey. Attempts to develop means for increasing tydeid mite numbers have not been successful, but cover crops that produce windblown pollen show promise. (Cover cropping and other cultural practices that help relieve spider mite problems are discussed below.)

Willamette mites, as noted, are much less serious pests of grapes than are Pacific and also act as an effective alternate prey for *M. occidentalis*. Along the San Joaquin Valley's eastern side and occasionally in north coast counties the twospotted mite is sometimes an important alternate prey on grapevines for *M. occidentalis*. Weedy vineyards are most likely to contain this spider mite species which produces only minor damage to grapes.

Other predators. Predatory insects and spiders are

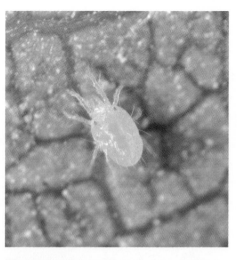

Willamette mite male waiting to mate with female as soon as she molts to the adult stage.

Willamette mite female and egg she has laid.

Diapausing form of Pacific mite female. Note lack of spotting.

Immature six spotted thrips attacking Pacific mite.

Orius "minute pirate bug" nymph feeding on mite.

Six spotted thrips feeding on mite.

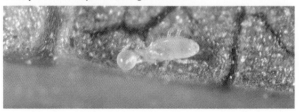

Typical early Pacific mite colonization injury in leaf fold.

Metaseiulus occidentalis attacking spider mite egg.

Tests show that more-frequent-than-usual sprinkler irrigations may be needed to control mites. For example, it was necessary to sprinkle approximately every 10 days for 12 hours during late June and July to keep Pacific mites in check in the west Fresno County area where this pest is severe. Also, sprinkling should be discontinued in Thompson Seedless vineyards by the end of July or early August to prevent berry cracking. Normally raisin growers do not irrigate later than July, so this should present them no problem.

Although permanent sprinklers are costly, they may be the most practical way to culture grapes in certain areas (e.g., west Fresno County) where the chances of correcting the imbalance of Pacific mite is now uncertain and the cost of repeated treatments may be prohibitive. Also, the cost of sprinklers becomes even less prohibitive when other factors, such as labor, water management and irrigation efficiency, are considered. Some growers with vineyards on sandy soil have reported significant increases in yields where overhead sprinklers have been installed.

Obviously, no suggestion offered here guarantees protection against spider mites. However, these practices, along with discriminate use of pesticides, can help reduce problems.

Evaluation of other vineyard pests. Treatment of grape leafhoppers may aggravate or cause spider mite problems. The decision to control leafhoppers should be based on established guidelines rather than on the calendar or preventive treatments. Treatment of first broods is believed more disruptive than treatment of summer broods. Some chemical applications for leafhoppers (carbaryl*) may actually stimulate reproduction of spider mites on grapes. If the same chemicals also reduce predaceous mites, then spider mites are given a dual advantage; delaying treatments gives predaceous mites an opportunity to increase without giving any advantage to Pacific mites.

If grape leafhopper populations can be tolerated until summer broods, treatment may then coincide with the need for treating spider mites. Chemicals are available that will control both leafhoppers and spider mites, some of which are less disruptive than others (see discussion on chemical selectivity). Other pests such as grape leaffolder and omnivorous leafroller are satisfactorily controlled without disturbing the natural control of spider mites. (See sections on these pests.)

Vineyard evaluations.

(1) Vineyard vigor and tolerance for spider mite damage needs careful evaluation. **Weak:** Vineyard shows little tolerance for Pacific mites; only light-to-moderate damage may be tolerated. **Average:** Vineyard is vigorous enough to tolerate some moderate-to-heavy damage. **Strong:** Vineyard can tolerate some heavy damage.

(2) Soil types and other related problems require careful evaluation. **Sandy soils:** Vines growing in sandy soils generally have lower tolerance for Pacific mites; during hot weather vines suffer moisture stress, despite frequent irrigations. **Restricted vine growth:** Alkaline soil, soil compaction, poor water penetration, nematodes and phylloxera all restrict vine growth and produce conditions favorable for Pacific mites. **Dusty conditions:** Dusty vineyards favor Pacific mites. Treatments are less effective on dusty vines. Frequent cultivation should be avoided.

Evaluating spider mite treatment levels. Thompson Seedless vineyards grown for raisin production in the San Joaquin Valley suffer the worst Pacific mite problems. Thus, most research on evaluating them has been in raisin grapes, but the principles outlined here help in evaluating treatment levels in other vineyards. It is important for pest control advisers and growers to recognize the value of alternate prey for predaceous mites. In the San Joaquin Valley, Willamette mites should be considered a beneficial alternate prey. In those areas where Willamette mite is found, a high population should be tolerated, except in the few table grape situations previously mentioned.

Evaluating and treating various Pacific mite situations. (See previous definitions of Pacific mite damage levels and predator distribution.)

(1) **Spot infestations.** If there is no evidence of a spread and the infested area is small, omit treatment. This allows predators to increase or spread (e.g., sixspotted thrips). If the infested area or areas are relatively large but constitute only a small part of the vineyard, spot treatments are suggested, providing predators are *rare* or *occasional* and leaves are beginning to burn. However, if predaceous mites are *frequent,* hold off treatment until it is obvious the predators cannot prevent too much damage. Use the most selective chemical available.

---

*Restricted material; permit required from County Agricultural Commissioner for possession or use.

(2) **General infestations:** If Pacific spider mite populations are *light* and predaceous mites are *rare* or *occasional*, delay treatment until weaker vines show some light bronzing, but not burning. If vines are vigorous, allow some bronzing of foliage. A little burn is permissible, but do not allow heavy burn or webbing on shoot tips. The holding action is necessary to increase predator numbers and improve their distribution. Remember: Well-distributed predator populations effectively prevent resurgence of Pacific mites after treatments.

If Pacific mite populations are *moderate* or *heavy* with predators *rare* or *occasional*, immediate treatment with the most effective acaricide is usually necessary. It may be possible to hold off treatment for a few days, if the vineyard is vigorous and amply supplied with moisture. Don't allow excessive webbing or burning to develop.

If Pacific mite populations are *moderate* to *heavy* with predaceous mites *frequent* to *numerous*, make careful evaluations of population trends. Is the spider mite population decreasing, stable or increasing? Is the distribution of the predator population improving? Inspect the vineyard frequently (twice a week or more); remember both predaceous and spider mites have enormous capacities to increase. Irrigate to avoid moisture stress. Do not cultivate or create dust. Let natural cover grow. Hold off treatment unless it is obvious damage is too great and predators are not yet effective. If vines are vigorous, bronzing and some burning should be tolerated. Again the holding action is to make as much use of predators as possible. All work has shown that treatments are much more effective if advantage is taken of predation.

Chemical selectivity. Fortunately, predaceous mites on grapes show considerable resistance to a number of organic pesticides, in particular the organic phosphates. Some chemicals reduce leafhopper and/or spider mite populations while allowing predaceous mites to maintain control of the latter. Parathion*, Ethion*, Trithion and Thiodan* have been used selectively in vineyards.

Sevin*, Cygon and Phosdrin* show little selectivity to predaceous mites on grapes. The selectivity of Dibrom, Guthion* and Zolone is little known. They appear less selective than Ethion and Parathion but more selective than Sevin, Cygon and Phosdrin. Guthion is widely used as a selective insecticide in deciduous fruit tree pest management.

The acaricides Kelthane and Omite show some interesting selective actions. Kelthane is not selective for predaceous mites, but it has little effect on sixspotted thrips. Thus, in situations where sixspotted thrips is the most important predator, Kelthane may be used to reduce Pacific mite populations without interfering with predation. Like Kelthane, Omite shows little effect on sixspotted thrips. On the other hand, Omite shows considerable promise as a selective acaricide for predaceous mites, particularly if the latter are allowed to increase before treatments are applied. However, detailed studies are needed to determine how various rates of Omite affect predator and prey populations in grapes. Ideally, treatments should be applied so that Pacific mites are reduced below economic levels without killing predaceous mites or reducing their food sources (Pacific mites and/or alternate prey) to the extent that they starve. This has been accomplished in other crops by using low rates of Omite and Plictran.

Studies indicate that methomyl* (Lannate and Nudrin) for leafhopper control is less disruptive if treatments are applied later in the season. Its short residual action allows predatory mite populations to recover. Early season treatments with methomyl (e.g., first brood of grape leafhopper) greatly hinder development of good predator populations. Methomyl should also be used only for late season control of omnivorous leafroller. Early season control for this worm pest is best accomplished with other materials (see section on omnivorous leafroller).

These studies also have revealed that at least one pyrethroid insecticide (permethrin, not registered for grapes at this time) has no selectivity for predaceous mites, even at very low application rates. More studies will be needed to determine whether the pyrethroids as a group are detrimental.

Less information is available on the effects of insecticides and miticides on tydeid mites. Detailed studies, however, have been accomplished with methomyl and one pyrethroid compound. In vineyard and laboratory tests, methomyl showed little or no effect on tydeids, while the tested pyrethroid was devastating, again even at very low rates. Preliminary studies indicate that Omite has only temporary effect on tydeid mite populations in vineyards.

Finally, it is important for both grower and pest control adviser to watch carefully for signs of selectivity and its prospects. Experience is just as

---

*Restricted material; permit required from County Agricultural Commissioner for possession or use.

important as experimentation in seeking out this important aspect of pest management. Pest managers should be continually cognizant of what treatments are doing to their beneficials as well as the pest they are attempting to control. Observations or population monitoring after treatment may pay dividends.

OMNIVOROUS LEAFROLLER

OLR female moth.

OLR egg mass on berry.

Since the 1960s the larva of the omnivorous leafroller (OLR), *Platynota stultana* Walshingham, has become a major pest widely distributed in California's San Joaquin and Sacramento valley vineyards. It has also been found in some Napa Valley vineyards.

The OLR can directly reduce grape yields by injuring the flowers and developing berries it feeds on and especially by allowing entry of bunch rot organisms that damage clusters. Leaf feeding is generally minor.

As the name "omnivorous" implies, the OLR larva is able to feed and develop on many different plants; therefore, adjacent orchards, row and field crops, ornamentals and weeds may be sources of infestation for the vineyard. Other crops attacked by OLR, besides grape, are: alfalfa, apple, apricot, avocado, bushberries, celery, citrus, cotton, lettuce, melons, peach, pepper, plum, prune, sorghum, sugar beet, strawberry, tomato and walnut. Susceptible ornamentals include: aster, carnation, chrysanthemum, cyclamen, eucalyptus, fuschia, geranium, portulaca and rose. Susceptible weeds are: pigweed, horseweed or mare's tail, panicled willow herb, cheeseweed, California mugwort and lambsquarters.

This tortricid moth was first described in 1884 from specimens collected from Sonora, Mexico. It was first found in California in 1913 on citrus nursery stock in Whittier. Since then the insect has been found on many plants, giving rise to many common names for it such as orange web worm, orange calyx worm, orange *Platynota*, rose leafroller, leaf tier, carnation moth and cotton leafroller. Besides California it has been found in Arizona, Florida, Illinois, Massachusetts, Michigan, Texas, Virginia and Washington, D.C., as well as in Mexico in Baja California, Sonora and Sinaloa. In midwestern and eastern states OLR is primarily a greenhouse pest.

Initial California OLR infestations on grapes were found in southern San Joaquin counties in the 1960s; several factors appeared to contribute to the problem. For example, extensive use of herbicides to control weeds in vine rows led to declining use of row plows; the subsequent buildup of residual trash, including mummified grape clusters where OLR larvae often overwinter, remained under vines. The larvae survived and in time built up in damaging numbers.

Additionally, some recently developed pesticides that control target pests, such as grape leafhopper, did not affect OLR but allowed the infestation to increase in the vineyard. Lastly, a change in the habits of OLR may have taken place, for in recent years OLR has become a problem not only on grape but on deciduous fruit trees, melons, sugar beets, cotton and seed alfalfa.

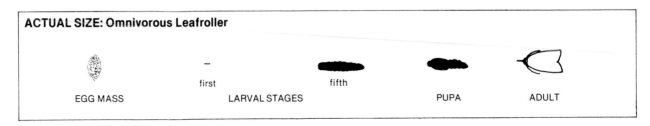

ACTUAL SIZE: Omnivorous Leafroller

EGG MASS   first   fifth   PUPA   ADULT

                  LARVAL STAGES

LARVAE COMMONLY FOUND IN GRAPE BUNCHES

Closeup showing white tubercles on mature OLR larva.

Orange tortrix larva lacks whitish tubercles in contrast to OLR larva.

Navel orangeworm showing crescent-shaped marking.

Grape leaffolder showing three markings on side of thorax.

Driedfruit beetle larva showing two small brown tubercles on end of abdomen.

Raisin moth larva showing rows of purple spots.

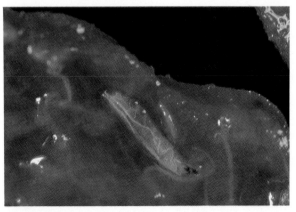

Drosophila larva with black mouth parts in narrow head region.

Early harvest. Where infestations occur in late summer, harvesting at the earliest possible date is recommended to avoid continued buildup of the OLR population and subsequent loss to rot. Early harvest is particularly important if raisins are going to be made in OLR-infested vineyards. With moderate OLR populations mold and rot problems can be reduced by picking before September 1. An early September harvest keeps more larvae from moving into the bunches where they are apt to continue to feed and make nests on the trays. Do not attempt to make raisins from heavily infested grapes because they are likely to be rejected.

Chemical control. Usually, spring treatments (bloom sprays) are recommended if the vineyard has a history of OLR problems, or if serious infestation occurred in the previous season. Otherwise, chemical treatments should be used only when determined necessary by monitoring,

Early season treatment with cryolite (Kryocide) spray has proved effective in reducing the OLR population and does not appear to aggravate such other pests as spider mites or leafhoppers. Where infestation is light to moderate, cryolite provides adequate seasonal control. Cryolite spray on table grapes is restricted to a single application made before fruit set. On wine and raisin grapes two applications can be made, the latter at least 30 days before harvest. On table grapes spray is generally applied at bloom, and on raisin and wine grapes shortly after bloom. At these periods clusters are exposed to the spray with minimal foliage interference. Most OLR larvae will be found in the clusters, so good coverage of clusters is critical; cryolite is a stomach poison and must be ingested by the insect to be effective.

Besides cryolite, several other insecticides can be used for later applications. These are: B.T. *(Bacillus thuringiensis)* preparations (Dipel, Thuricide and Biotrol XK), carbaryl* (Sevin), methomyl* (Lannate and Nudrin), phosmet (Imidan) and phosalone (Zolone).

B.T. may be used in vineyards with light OLR infestations and will not reduce the beneficial predatory mite population if Pacific spider mites are a problem in the vineyard.

Carbaryl and methomyl increase Pacific spider mite infestation in some vineyards. If these chemicals are used for OLR control, they should be applied for late season control or in areas where mites are not a problem.

---

*Restricted material; permit required from County Agricultural Commissioner for possession or use.

The variegated grape leafhopper, *Erythroneura variabilis* Beamer, is the principal vineyard pest and the only leafhopper species found in large numbers in southern California. Infestations are especially severe in the Coachella Valley because high temperatures allow the pest's rapid development and favorable overwintering quarters are abundant. To survive, overwintering adult leafhoppers must have access to green vegetation.

Description

Mature variegated grape leafhoppers are about 3 mm (1/8 inch) long, richly mottled in brown, green, red, white and yellow. Newly hatched nymphs are white; mature nymphs are yellow to yellow-brown with wing pads.

Injury

Nymphs cause the principal damage, black spotting of table grapes, which detracts from their market value. Spotting results when sooty mold develops on excreted honeydew. When adults are numerous, they interfere with harvesting by annoying the pickers. When control is neglected, high populations will cause partial to complete defoliation, thereby weakening the vines.

Seasonal Development

The population dynamics of the variegated grape leafhopper varies in southern California's two major grape growing areas, the Coachella Valley and the San Bernardino County area.

Coachella Valley. Large numbers of adult leafhoppers overwinter in windbreaks of tamarisk, bamboo and eucalyptus, in citrus orchards, in hedgerows of pyracantha and oleander, in grain fields and on desert plants such as sagebrush. They are especially numerous in bermudagrass in date gardens. Movement into vineyards begins soon after vine growth starts in spring. (See table.)

At mean temperatures of 32°C (90°F), the variegated grape leafhopper completes one generation in 23 days. Because there are six generations each year and each

---

ACTUAL SIZE: Variegated Grape Leafhopper of Southern California

EGG	first NYMPHAL STAGES fifth	ADULT

Typical Early Season Development and Treatment Times for Variegated Grape Leafhopper in Coachella Valley.*

First movement of overwintering adult leafhoppers into vineyards	March 7–15
First oviposition	March 21–30
Best time for treating vineyard and adjacent windbreaks for overwintering adults	April 1–15
First generation egg hatch begins	April 7–15
First generation nymphs	April 7–May 25
First generation adults appear	May 10
Second opportunity for preharvest control †	May 15–25
First generation egg hatch practically complete	May 25
First generation adults at high population level	June 1–10
Second generation nymphs present in large numbers and spotting fruit	June 1–25

*Seasonal development will vary from year to year; therefore, dates given are intended only as a guide.
†No treatment required if overwintering adults have been adequately controlled.

Variegated leafhopper adult.

First instar nymph of variegated leafhopper.

Fifth instar of variegated leafhopper.

Third instar of variegated leafhopper.

female can lay as many as 75 eggs, extremely high populations are not unusual in Coachella Valley vineyards.

San Bernardino County area. Adult leafhoppers overwinter on almost any green vegetation adjacent to or in the vineyard. After winter rains they commonly are found on foxtail, foxtail chess, clovers and especially filaree, and from late March to mid-April they move into vineyards. The first eggs are laid about three to four weeks later in mid-April to mid-May. Leafhoppers develop more slowly in the San Bernardino area; there are four generations annually compared with six in the Coachella Valley.

Natural Control

Eggs of the variegated grape leafhopper, like those of its northern California counterpart, the grape leafhopper (see page 100), are parasitized by *Anagrus epos* (Girault). However, no parasites attack this leafhopper in the Coachella Valley, where the natural blackberry refuges for *Anagrus* are absent.

On the other hand, *Anagrus* is active in the San Bernardino area, according to studies in a vineyard at the mouth of Lytle Creek Canyon, about six miles from wild grapes and blackberries in the canyon.

Anagrus parasitization of leafhopper eggs occurs there in late May or early June, about 40 days after variegated grape leafhopper eggs appear in spring. By August parasitization has reached 90 percent. Later in the season parasites appear in vineyards near Ontario airport and along Interstate 10 with less impact on leafhopper vineyard populations. However, the peak average level of 72 percent parasitism in late September does reduce the overwintering population.

For more detailed information on the biology of *Anagrus,* see the section on grape leafhopper, page 100 *(Natural Control).*

Monitoring Guidelines

Overwintered adults are best surveyed before vines begin to leaf out, especially on warm days when they are active. About three to four weeks after these adults move into the vineyard, first generation nymphs are usually found on the lower surface of the basal six to eight leaves. Few, if any, nymphs are found on younger leaves further out on the shoots, so do not be misled about population status because of lack of nymphs in these areas.

Procedures have not been developed for monitoring later nymphal generations. To assure accuracy of vineyard counts, anyone checking these later populations should determine first which leaves carry the highest nymphal levels.

Management Guidelines

Systematic assessment of economic levels (e.g., numbers of nymphs per leaf) has not been developed for variegated grape leafhopper. However, a clear understanding of its seasonal development in the two southern California areas will promote good management decisions.

Coachella Valley. First opportunity for preharvest control: Where large populations of overwintering leafhopper adults can be expected to invade annually in early spring (March 7–15), a control application should be applied when egg laying begins (April 1–15) or about 14 days after the first leafhopper invasion. At that time, because only part of the population has moved into the vineyard from overwintering quarters, the first vineyard treatment should be extended outside to all adjacent noncrop, hedgerow vegetation. This is the optimum time for control on all varieties (Thompson Seedless, Perlette, Cardinal), as visible insecticide residues on the fruit must be avoided at harvest.

If good control is obtained when egg laying begins and if the vineyard is not subject to gross infestation from adjacent untreated vineyards, an additional treatment should not be required until the postharvest period. If this is not the case, another opportunity for treatment occurs, as outlined below.

Second opportunity for preharvest control: Eggs producing the first generation begin to hatch 18 to 21 days after egg laying begins (April 7–15) and continue to hatch for about six or seven weeks. But before this hatch is completed, some of its earliest adults will have matured and laid eggs. Therefore, a second op-

portunity for good control comes when this first generation hatch is almost complete and the overlapping second generation hatch is beginning. In the Coachella Valley this usually occurs May 15–25. (This second opportunity for timing a treatment is less suitable for early varieties such as Perlette and Cardinal, because it may overlap harvest.) At this time the population of adult leafhoppers is relatively low (overwintering adults have died and development of new adults has just begun). The number of mature first generation nymphs will be high, however, and if left uncontrolled, they will cause spotting of early varieties. The second generation they could produce would then become numerous in the vineyard just before and during Thompson Seedless harvest, creating additional spotting of grapes.

For spraying after fruit set, a wettable powder formulation should be used instead of an emulsifiable solution, as the latter may scar the berries.

One postharvest treatment for leafhoppers is generally required to prevent midsummer defoliation of vineyards. Three to four generations develop after harvest.

San Bernardino County area. Oviposition by overwintered adults begins about 20 days after they are first observed in the vineyard in early spring. If treatment is effectively applied at that time and there is no heavy reinfestation from adjacent untreated vineyards, no further treatment should be required. This time is well before most local vineyards are invaded by leafhopper egg parasites.

Vineyards distant from foothills or without favorable nearby leafhopper overwintering quarters do not require treatment for leafhoppers in some seasons in this area.

In parts of California the larvae of the western grapeleaf skeletonizer (GLS), *Harrisina brillians* Barnes & McDunnough, is a serious defoliating pest of vineyards, backyard grapevines and wild grapes in parks and along rivers and streams. Under certain conditions it feeds on the fruit, and then bunch rot usually destroys the entire cluster. The larvae, especially larger ones, can sting and cause skin welts on field workers who touch them.

Ornamental hosts in the grape family are Virginia creeper and Boston ivy. This pest also has been observed feeding on apricots, almonds, cherries and roses, but these are presently considered incidental hosts.

The skeletonizer was originally distributed through Arizona, New Mexico, Utah, Colorado, Nevada and Mexico's Sonora, Chihuahua, Coahuila and Aguascalientes states. It was first found in 1941 in California, near San Diego. In canyon areas wild grapes, *Vitis californica*, were severely defoliated, and in a short time GLS became a serious pest in commercial vineyards. Under a quarantine established in 1942 movement of untreated fruit was prohibited; methyl bromide treatment of fruit containers was required to prevent its spread. By 1943 crop loss in some vineyards reached 90 percent (the average was 40 to 60 percent).

A state-imposed eradication program, using cryolite dust, was tried, but it was not successful. Meanwhile, emphasis was placed on biological control. U.C. entomologists at Riverside imported parasites from Mexico and Arizona; a parasitic wasp, *Apanteles harrisinae* Muesbeck, and a parasitic fly, *Ametadoria* (=*Sturmia*) *harrisinae* (Coquillett), were soon established. A granulosis virus that attacks GLS lar-

vae was found in nature in San Diego County and in Arizona, and also in GLS rearing laboratories. The virus is still an important bottleneck in the mass-rearing of GLS for production of parasites for release. Granulosis is spread by the parasites and may exert considerable control on GLS field populations.

By 1961 GLS was found on backyard grapes at Kerman (Fresno County). In spite of another try at eradication, by 1975 infestations could be found in central and northern California on wild grapes and in backyard and commercial vineyards. The pest has apparently been eradicated in Lake County, but it has not been seen in north coast or central coast districts. It has spread slowly in the Central Valley because the adults tend to remain in the area of their larval development.

To deal with these new infestations *Apanteles* and *Ametadoria* (=*Sturmia*) parasites from San Diego County were introduced and explorations made for new parasites from outside of California. So far only the parasitic fly, *Ametadoria harrisinae*, appears to have been established and this only in Siskiyou and Shasta counties where grapes are not commercially important.

Description

GLS adults are conspicuous because of their bluish-black to greenish-black color. Normal body length is about 15.9 mm (5/8 inch) and the wingspan is about 25.4 to 33.9 mm (1 to 1-1/3 inches). The adults are much smaller, however, when food is short for developing larvae.

Both sexes have comblike bristles projecting from the antennae. Male bristles, about 3/4 mm long, are twice

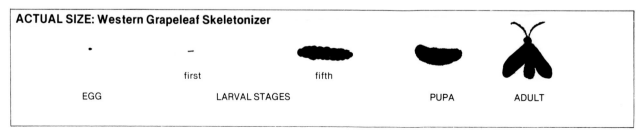

ACTUAL SIZE: Western Grapeleaf Skeletonizer

EGG   LARVAL STAGES   PUPA   ADULT

first   fifth

as long as those of the female. The male also has fine hairs on each bristle; the female does not. The male has posterior tufts of hairs on the tip of the abdomen; these are absent on the female. The posterior end of the female abdomen is more or less squared off, while the tip of the male's posterior is more pointed.

GLS capsular-shaped eggs are about 0.4 mm wide and 0.62 mm long. They are usually pale yellow, but some batches are whitish. All are ornamented with fine surface reticulations. Eggs are laid on a horizontal plane in relation to the leaf surface in clustered groups (although the eggs do not touch one another). There may be seven or eight or as many as 300 or more in a cluster—the average is 96. Often there may be two or more clusters per leaf, but these are not usually laid by the same female.

Nearly 100 percent of the eggs are laid on the undersides of leaves. In commercial vineyards eggs of the first generation are laid mostly on the inner mature leaves under the vine canopy. Eggs of the second and third generation are almost always laid on the leaves of the outside shoots. This makes it easy to look for egg batches by lifting the shoots and looking for eggs on the underside of leaves.

All eggs in a cluster hatch at about the same time. The first of five instars is pale white and about 1.6 mm (1/16 inch) long. These larvae line up side by side in a circular row to feed. They strip the tissue of the underside of the leaf to about the size of a half dollar or less, depending on their number. On the upper surface the feeding area appears as a white spot.

Strongly gregarious, the larvae habitually feed side by side in the first three instars and sometimes in the early fourth instar. When the first three instars are ready to molt, they move away from their feeding site, form a round conglomerate mass and then molt. Often they may move a short distance to an undamaged leaf and sometimes to the stem and shoot to molt. Recognition of this habit is important and can be deceiving when only sampling damaged leaves for young larval colonies.

The second instar larvae are about 3.2 to 4.8 mm (1/8 to 3/16 inch) long. They are yellowish-white with black hairs on the body segments. The hairs are commonly dotted with pellets of black excrement. Late in this stage, two large pale brown rings appear on the body.

The third instar is about 6.4 to 7.7 mm (1/4 to 5/16 inch) long. This stage takes on all of the permanent body-ring colorations, consisting of two large brown bands that mark the body into thirds, a forward portion, a middle section and the posterior. There are five narrow brown bands: two around the forward portion, one on the middle section between the large brown bands and two on the posterior. Body color between the bands is pale brown.

The fourth instar is about 11 mm (7/16 inch) long and the fifth instar about 15.9 mm (5/8 inch) long just before pupation. The seven circular bands on the body become blackish-purple in the fourth and fifth instar; body color between the bands turns bright yellow.

Fully grown larvae spin silken cocoons in which to pupate. The cocoons, irregular, dirty-white capsules, may be found in trash around the base of vines or under loose bark on the trunk.

Injury

Leaf damage continues to increase through the season in untreated populations. The white feeding spots of the early instars on leaves merge as the larval groups feed and move during their developmental stages.

The larvae feed in conspicuous, orderly rows in the first to third instars and sometimes in the early fourth instar. The first to third instars and many early fourth instars nearly always feed on the leaf underside, leaving only the veins and upper cuticle intact. These white, damaged leaves are easily seen in surveys of GLS infestations. The gregarious feeding habit is less evident in the fourth and fifth instars. A few larvae will skeletonize an entire leaf, leaving only the larger veins.

Second and third generations (there are three per year) have defoliated entire vineyards in the central San Joaquin Valley, and larval feeding has been observed into November. If vines are partially or completely defoliated before harvest, fruit maturity can be adversely affected. Heavy defoliation also can cause fruit sunburn and quality loss. Defoliation after harvest is less damaging, but it may affect the food reserves of the vine and weaken it for the following year.

Male and female moths copulating. Larger moth on right is the male.

Three typical egg masses laid on the underside of grape leaf.

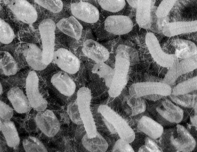

Closeup of skeletonizer eggs, upper photo, shows their typical barrel shape. In lower photo, first instar larvae are hatching.

First instar larvae lining up side by side to feed in circular formation.

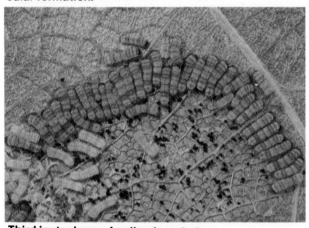

Third instar larvae feeding in a circle.

Fifth instar larvae are much less gregarious.

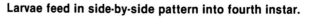

Larvae feed in side-by-side pattern into fourth instar.

Skeletonizer pupating under the bark in cocoons. Inset shows closeup of details of typical cocoons.

Typical first generation damage is seen in top photo on lower shaded leaves of vine. Photo below shows typical whitish spots caused by feeding damage on lower surface.

Ordinarily GLS does not feed on fruit. However, where there are high population levels, fourth and fifth instar stages will feed on fruit once they have consumed most of the foliage. Their damage to the berry skin allows bunch rot to set in, and this usually leads to loss of an entire cluster.

Many field workers are sensitive to contact with GLS larvae. There are many long, dark hairs on the larval body—easily seen on the fourth and fifth instars—and these seem to be poisonous. When workers brush against the larvae, skin welts similar to those produced by nettles can result. Some people have a

greater reaction than others—some even exhibit a skin reaction to the hairs of the second and third instar.

Seasonal Development

GLS overwinters in the pupal stage. Because the threshold for grape plant development is 10°C (50°F) and for GLS 15.5°C (60°F), emergence, mating and oviposition of first generation adults do not occur until after the flush of new shoots and leaves on grapevines in spring.

The female prefers shady areas of the vine to lay eggs and rarely lays on developing shoot growth. Egg laying occurs one to four days after mating, although in spring it may be delayed slightly longer because cool weather slows egg development in the female. Adults of the first or spring generation live four to eight days, compared with three to five days for adults of summer generations.

Larvae have five instar stages before pupating into adults. A small portion of the first and second generations go into a pupal diapause; essentially all larvae pupating in September and thereafter enter diapause. Overall population levels increase with each succeeding generation.

In San Diego County first generation adults emerge from diapaused pupae from late April through mid-May. The eggs of this generation hatch in an average of 16 days or from about early through late May. The larvae evolve into second generation adults that emerge from about mid- to late July. Eggs of this second generation hatch about 10 days after oviposition or from late July into early August. Most larvae that complete development go into diapause; a few initiate the partial third generation.

In the San Joaquin Valley there are three complete generations per year. All larvae of the third generation enter diapause and emerge the following spring. The appearance of adults and egg laying occur first in urban areas, then in commercial vineyards and about two to three weeks later in wild grapes. The table below shows the sequence for the three generations in San Joaquin Valley commercial vineyards.

The developmental activity of each generation occurs over distinct periods of time in vineyards removed from wild grapes, but there can be an overlap of the second and third generation when vineyards are near wild grapes. This occurs when GLS larvae have severely defoliated wild grapes and the adults disperse to find new egg laying places.

Natural Control

In San Diego County biological control of GLS by the imported *Apanteles* and *Ametadoria* parasites, plus the natural granulosis virus disease, has markedly suppressed GLS. However, damage sometimes occurs in backyard vines and in commercial vineyards.

The parasitic wasp, *Apanteles harrisinae* (Braconidae), attacks early instars of GLS and the parasitic fly, *Ametadoria harrisinae* (Tachinidae), usually attacks older instars. Both can also transmit the granulosis virus, which can effectively lower GLS population levels. Failure of egg hatch and diseased larvae have been observed in Tulare County, and evidence suggests that the virus may be present there. The parasites have not become established in the Central Valley, although efforts to establish them have been made and are continuing.

Studies in San Diego County have shown that *Ametadoria* is affected little, if at all, by sulfur dust, whereas *Apanteles* suffers severely. Little information is available on the effect of insecticides on GLS parasites.

Monitoring Guidelines

The present monitoring method for adults is to count the number of GLS observed flying between two stakes, 25 feet apart, between two vine rows in one minute. The 25-foot counting zones should be established in several areas of the vineyard where the pest is present or suspected. A count of 20 or more adults flying through the 25-foot zones in one minute in all parts of the vineyard indicates there will be severe defoliation and 100 percent fruit loss. Counts of six to seven adults per minute per 25 feet indicate that

Occurrence Periods and Stages of Western Grapeleaf Skeletonizer in Central San Joaquin Valley Commercial Vineyards.*			
	Adults	Eggs	Larvae
First generation	Late April to mid-May	Late April to mid-late May	Early May to late June
Second generation	Late June to early July	Late June to mid-July	July to mid-August
Third generation	Late August to early September	Late August to mid-September	September and October
*Data were collected in Tulare County. Time periods may change slightly in warmer or cooler areas.			

widespread leaf damage will occur. Counts should be made from midmorning to early afternoon.

Adults fly in the daytime and usually near the ground between vine rows. Greatest flight activity is in the morning; it decreases considerably by mid-afternoon. Nearly all flight ceases by late afternoon, at which time adults will be mainly inside the vine resting on the trunk, arms and shoots. Cloudy conditions, it has been noted, reduce flight activity.

In the San Joaquin Valley first generation eggs are seen in late April to mid-May, mostly on the mature basal leaves under the canopy. Second and third generations are mainly on shoots hanging close to the ground. Early instar colonies are seldom found in the vines' upper parts; fourth and fifth instar larvae are commonly found there.

Small, circular whitish spots on the upper surface of a leaf indicate the presence of newly developing colonies on the lower surface. Later, as the developing colony strips off all of the lower epidermis, the entire leaf becomes paperlike.

New infestations usually begin on the end vines and along border rows, and these areas are likely to have the highest GLS populations. This results from adults migrating in after defoliating vines elsewhere. The infestation then spreads through the vineyard in the following generations because the adults tend to remain in the area of their larval development. This also accounts for the remarkably slow spread of GLS in the Central Valley.

Vineyards should be monitored closely for leaf damage and defoliation just before harvest. The "stinging" ability of the hairs on GLS larvae also should not be taken lightly; even moderate populations of mature larvae may pose hazards to pickers.

Management Guidelines

Present biological control is inadequate and GLS populations are generally free to develop unchecked, except in San Diego County. High larval populations in the fourth and fifth instar can readily defoliate a vineyard, unless an effective insecticide is used. Serious infestations may develop in untreated vineyards or on backyard vines; these should be examined for eggs and larvae during the season. Spot treatments will suffice if the vineyard infestation is localized. Sprays appear to give better control than do dusts.

The skeletonizer is easy to control with chemicals used for lepidopterous larvae on grapes. A key feature of GLS chemical control is good underleaf coverage because the first to third and many early fourth instars feed almost entirely on leaf undersides. One correctly timed application that gives good larval cleanup will usually provide season-long protection from GLS, except in vineyards near wild grapes. Such vineyards may be vulnerable to reinvasion after the wild grapes have been defoliated. Invasion of adults from neglected or mismanaged vineyards can also threaten neighboring vineyards.

Chemical control with an insecticide such as cryolite (Kryocide), a stomach poison, is preferred. It has a long residual action and, unlike many contact insecticides, it is less harmful to natural enemies. A GLS treatment can be timed with control measures for the omnivorous leafroller.

Naled (Dibrom) and methomyl* (Lannate, Nudrin) have a short residual life and have been used for late-season leafhopper control on table grapes (they will also eliminate the GLS that can sting field workers). Carbaryl* (Sevin) gives satisfactory GLS control but its residual life is not as long as cryolite. Treatment should be applied after all eggs have hatched. Sevin may also cause spider mite problems. *Bacillus thuringiensis* (Dipel, Biotrol XK, Thuricide) gives fairly good control of GLS and is not harmful to natural enemies; it has a short residual life. If coverage is not satisfactory or if all eggs have not hatched, a second treatment may be necessary.

Larvae have to ingest a sufficient amount of cryolite (Kryocide) and B.T. before being killed by them, although GLS leaf feeding is reduced markedly before the larvae die. Sick larvae usually fall from the vine and die five to eight days after treatment.

The following spring vineyards with an uncontrolled GLS fall population will have the problem even more widespread and devastating than in the previous year.

---

*Restricted material; permit required from County Agricultural Commissioner for possession or use.

White damaged leaves typical of GLS first, third and many early fourth instar feedings are seen in the vineyard and closeup. These instars nearly always feed on the leaf underside, leaving veins and upper cuticle intact.

Fifth instar larvae skeletonizing entire leaf leaving only the larger veins.

Ametadoria harrisinae, a parasitic fly, ready to attack fifth instar. Inset shows detail of *A. harrisinae.*

Closeup of GLS fifth instar stinging hairs.

Apanteles harrisinae, a parasitic wasp of early GLS instars.

Grape leaffolder moth.

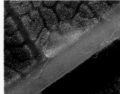

Grape leaffolder egg.

Grape leaffolder larva.

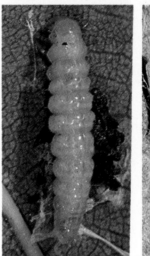

Prepupal stage is left. Right is pupal stage in opened leaf envelope.

The grape leaffolder, *Desmia funeralis* (Hübner), is a grape pest in the central and southern San Joaquin Valley, the degree of its infestations varying greatly from year to year. In the worst years less than half of the vineyards have economic infestations; with light infestation only a few leaf rolls are found.

The grape leaffolder (GLF) causes injury in the larval stage by rolling and feeding on the leaves, reducing photosynthetic function. Under extreme population densities it may feed on fruit, but economic damage usually occurs only with massive, late season infestations.

In recent years little fruit loss has occurred because of effective control measures. The appearance of leaf rolls makes the presence of this insect highly visible and this aids in monitoring population density.

GLF, apparently native to the east coast, reached California in the late 19th century and has few natural enemies here. It now occurs on wild grapes from coast to coast and is especially abundant in the southwest, southeast and Atlantic coastal regions north into Canada. It also ranges south to northeastern Mexico. Most, but not all, populations outside of California fold rather than roll leaves; thus, the common name grape leaffolder. The California population must have originated with one of the leaf rolling strains. Other plants recorded as host for GLF are evening primrose, Virginia creeper and redbud.

There are great annual fluctuations in GLF populations. In Tulare and Fresno counties heavy late season outbreaks occurred in 1945 and 1952 and annually from 1954 through 1959. The population then remained at a low ebb until 1964, when it then started a slow, somewhat inconsistent upward trend. Moderate populations have been encountered through the late 1970s.

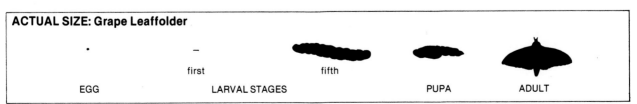

ACTUAL SIZE: Grape Leaffolder

first   fifth

EGG   LARVAL STAGES   PUPA   ADULT

Reasons for these fluctuations are not known, nor is it known why populations, even when low, remain in the same localized areas. In Tulare County, for instance, an area southwest of Exeter generally associated with the Kaweah River complex always has some GLF in vineyards as does an area in Fresno County along the Kings River between Centerville and Kingsburg.

Description

Coloration of the GLF adult is distinctive. Adults are black with two white spots on the fore wing. Males have one large irregularly shaped spot on the hind wings, while females have two smaller spots partly or completely divided; wing expanse is about 25.4 mm (one inch). Both sexes have varying amounts of white on the fringes of the wing and two white bands across the abdomen. Male antennae are elbowed in the middle; female antennae are smooth. The male abdomen is pointed; that of the female is blunt.

The eggs are small, flat, iridescent, elliptical and about 0.8 mm (1/32 inch) long. They are laid, usually singly, on either the upper or lower surfaces of leaves, with the lower surface preferred. Many eggs are laid against a vein. With large deposits the eggs may overlap. The easiest way to see the eggs is to remove a leaf from a vine and turn it slightly in the sun so that the eggs will reflect light. Both hatched and unhatched eggs glisten. The unhatched eggs are iridescent, while the hatched eggs look more silvery. Viewed with a hand lens, an unhatched egg is convex, while a hatched egg is concave with a torn edge where the larva has emerged.

The larvae are about 1.6 mm (1/16 inch) long when hatched and 15.4 mm to 22.2 mm (5/8 to 7/8 inch) long when fully grown. They undergo five instars before pupation. There are no characteristic marks for field identification on the first two instars. On the third instar a small black spot appears on each side of the body above the second pair of legs on the middle segment of the thorax, which is the division of the body just behind the head. On the fourth instar there are two spots, the second being indistinct early in this stage. The fifth instar has three distinct spots on the thorax and a spot appears near the anal end.

As soon as feeding begins, the ingested leaf tissue shows through the translucent body wall, giving the larvae a green internal color. Feeding stops at the end of each instar, the gut is voided and the green colora-

tion changes to a light brownish-tan during molting. Many persons have misinterpreted this coloring as evidence of disease, parasitism or insecticide poisoning. However, the typical green body color reappears with resumption of leaf feeding. Larvae feeding on grape berries do not have the bright green body of those feeding on leaves, but are rather uniformly light colored.

The pupal case is brown, about 12.7 mm (1/2 inch) long.

Injury

The extent of the vine damage through reduction of leaf surface by rolling and larval feeding depends upon population size and the time of year or brood involved.

Defoliation studies have shown that Thompson Seedless vines will tolerate about 20 percent leaf loss a month after fruit set and higher levels later. Vine tolerance to numbers of leaf rolls has not been studied in detail. Observations on Emperors indicate that first brood larvae (April to mid-June) usually are too low in numbers to do any damage and that this variety will tolerate about 200 leaf rolls per vine in either second (June to late July) or third broods (late August to late September) without adverse effect on crop. Healthy Emperor vines keep growing late in the season, so that leaf surface is partially restored from that lost to the second brood.

Second brood, however, is sometimes heavy enough to cause as much as 50 percent defoliation—well beyond an economic level—and a few worms may be found in the fruit. In heavy infestations the third brood developing from late August to late September can completely defoliate vines. In such a situation some larvae begin feeding on the fruit early and then, as available leaves decline, more and more larvae move into the fruit. Several kinds of damage can occur. With heavy defoliation fruit becomes soft and unfit for fresh fruit shipment. Or, if fruit contains large numbers of larvae, it may be rejected at the winery or diverted for distilling material.

Table grapes face an additional problem. Larvae ready to make their first leaf rolls tend to move to the top of the vine, a move especially noticeable in the second and third broods. This can result in a vine fairly well defoliated on top with fruit exposed to discoloration and sunburning.

In some varieties the third brood may defoliate just at or after harvest. New growth will then develop from lateral buds, which are not the primary buds that will give rise to the next year's crop. It is not obvious that any crop reduction results. Studies with Thompson Seedless raisin grapes show about a 10 percent reduction in crop the next year when vines are 60 percent defoliated in early September. The reduction is due to smaller, not fewer, clusters. Crop reductions do not become more severe than 10 percent, even when vines were defoliated for 10 consecutive years.

Grape leaffolder prefers native American grape varieties. Thus, rootstock vines, with American variety parentage, will be more heavily attacked than vinifera varieties.

Seasonal Development

There are three broods per year and sometimes a partial fourth. The average moth flight periods and time required for development stages are shown in the table below. Although there are three definite broods, there is some overlapping.

The figure on page 152 shows moth trapping studies in a four-year period. A long flight period for the first brood adults, a relatively short flight period for the second brood, and a slightly longer period usually for the third brood were observed. A few moths were in flight between the peaks of the brood. Generally, more moths were trapped in the second and third broods than in the first; 1942 was a notable exception.

Time of the broods may vary up to two weeks from year to year and up to 10 days from vineyard to vineyard in the same year. In any particular vineyard getting oriented as to the stage of the brood may take a little time, especially if the population is low.

Adults fly primarily at night during mating and egg laying. With light populations, only an occasional moth will be seen during the day because most remain hidden within the vine canopy. However, with heavy second and third brood populations, more adults will be seen flying around the vines during the day. Adults live about nine days in the field.

First brood adults seek wind protection. Thus, they may lay their eggs on vines around barns or windbreaks or seek the densest parts of the vine. Shoots developing near the ground surface are frequently preferred as egg-laying sites.

Females deposit most eggs on the second through fifth day of laying. Number of eggs per female has not been accurately verified; one worker reported 60 and another reported about 200 per female.

For first brood egg number per leaf is usually low. Second and third broods may have numerous eggs, as many as 10 to 15 per leaf with heavier broods. Most eggs are fertile and hatch in 10 to 17 days in the spring and four to five days in the summer.

The larvae undergo five instars before pupation. The young larvae usually feed in groups where numerous eggs are laid on one leaf. They seek protection between two touching leaves, between the overlapping area at the base of the leaf or, during the second and third broods, inside leaf rolls made by the previous brood. A very few, especially in the first brood, feed unprotected on the leaf surface. The young larvae

Brood	Egg Laying Period	Time Required for Eggs to Hatch	Total Time in Larval Stages	Total Time in Pupal Stage	Total Time, Egg to Moth Emergence
First	April 2 to May 24	10 to 17 days	3 to 4 weeks	10 days to 2 weeks	6½ to 7½ weeks
Second	June 15 to July 15	4 to 5 days	2 to 3 weeks	7 to 11 days	4 to 5 weeks
Third	Aug. 3 to Sept. 5	4 to 5 days	3 to 5 weeks	overwinter	

Average Moth Flight Periods and Time Required for Completion of Various Developmental Stages of the Grape Leaffolder (Fresno).

Photo by Frederik L. Jensen

Leaf rolls made by grape leaffolder. Inset shows closeup of grape leaffolder roll.

Feeding damage of leaves tied together by leaffolder larvae.

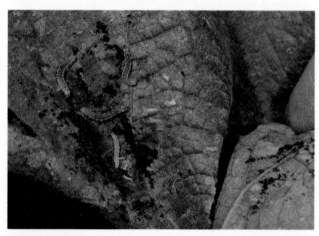

Leaves pulled apart showing young larvae and damage.

Very heavy grape leaffolder damage exposing fruit.

Bracon cushmani parasite stinging grape leaffolder larva.

Bracon cushmani eggs laid externally on grape leaffolder larva.

Bracon cushmani larvae feeding externally on grape leaffolder larva.

Bracon cushmani cocoons next to consumed grape leaffolder larva.

group feed on the leaf tissue but never completely eat through the leaf. They always leave the outer layers of epidermal leaf cells for protection, creating a characteristic pattern of feeding injury. Thus, the presence or past occurrence of young larvae, either between leaves or in old rolls, is easily recognizable. The epidermal leaf cells are colorless as they contain no chlorophyll. These cells then dry up and turn brown in one to three days. The tissue remains intact and continues to give protection to the larvae. With heavy populations there may be more than 20 larvae sharing a niche between touching leaves or in old rolls.

After about 10 days of feeding (second and third broods), the larvae are in the fourth instar and ready to make the pencil-sized leaf rolls. Some third instar larvae make small leaf rolls. The larvae make leaf rolls only during darkness.

Fourth instar larvae eat only the free edge of the leaf inside the roll. Usually there is only one larva in a new roll, although with very heavy broods there may sometimes be two. Shortly after a brood starts making and inhabiting leaf rolls, about half the rolls will be empty. This is because the larva exhausts its food supply inside the roll. It then vacates and rolls another leaf or pupates if fully grown. Each larva makes at least two leaf rolls before completing feeding.

Disturbed larvae wriggle vigorously to escape, frequently falling to the ground. They do not usually descend on a silken strand as does the omnivorous leafroller, *Platynota stultana* (Walsingham). Both species are sometimes found feeding together between leaves or in leaf rolls.

As the fifth instar larva completes development, it makes a pupal envelope on the edge of a leaf. The envelope is a small section of leaf partially cut away, folded and tightly webbed together from within. The envelope usually remains attached to the leaf in the first and second broods. The third brood envelope often separates from the leaf and is found underneath the vine with leaves and other debris.

Inside the envelope the larva becomes shorter and broader, turns pinkish and gradually transforms into a pupa. The pupal stage lasts from one to two weeks, except for the third brood pupae which overwinter. A few third brood larvae pupate underneath the grapevines' loose bark, and some first and second broods pupate in leaf rolls rather than in leaf envelopes.

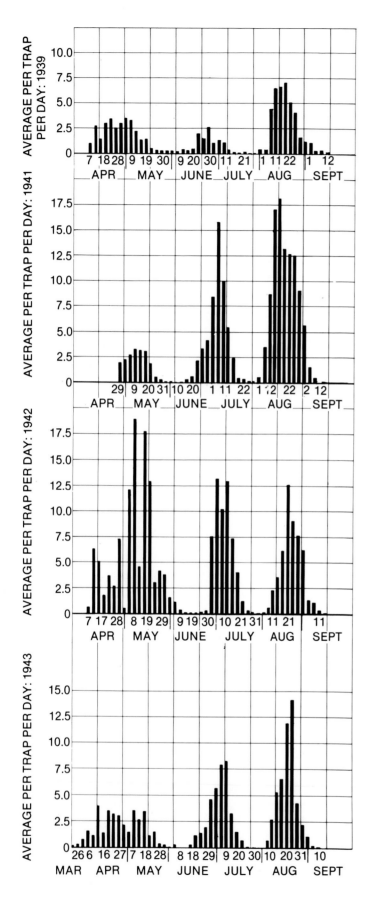

Natural Control

The most commonly observed parasite of the GLF is the larval parasite, *Bracon cushmani* (Mues.). This small wasp (hymenoptera), about the size of the gnat but with a fat abdomen, reproduces on larvae of about the third instar and older. After stinging and paralyzing the larvae, it lays one to several eggs on the body. The hatched wasp larvae feed externally on the GLF larva and pupate near the consumed body. Parasites per host average about eight but range from one to 20, depending considerably upon the size of the larva parasitized. Parasitism ordinarily is in the range of 30 to 40 percent but often higher. Some years these wasps may be seen flying around the vine by the hundreds.

Bracon cushmani usually increases in summer and frequently reduces the size of the second and third brood to such small numbers that little increase in GLF population is detectable.

Two tachinid parasite flies, (Diptera) *Nemorilla pyste* (Walker) and *Erynnia tortricis* (Coquillett), attack GLF. Both are larval parasites. *Nemorilla* is effective on later instars but appears too late to lessen damage by the worms. *Erynnia* attacks a variety of small lepidoptera. The females deposit one to several chalky or pearly white eggs around the GLF head and thorax regions. On hatching the parasitic larva invades the host body, develops and then pupates within the GLF pupa, causing it to appear shorter and more blunt than normally.

Two hymenoptera pupal parasites are occasionally collected, *Brachymeria ovata* (Say) and *Coccygomimus sanguinipes* (Cress.), but little is known of their activity or benefit.

GLF eggs attacked by *Trichogramma* sp. turn dark, but *Trichogramma* is not believed an effective parasite because of its erratic activity and the low percentage of parasitism. Efforts to introduce other wasp or beetle parasites and predators have been unsuccessful, including *Macrocentrus nuperus* Cress., and *Apanteles carnarsiae* Ashm., braconids; *Casinaria infesta* (Cress.), an ichneumonid; *Prosierola bicarinta* (Brues.),

←

Average number of grape leaffolder moths taken per trap per day during four years of trapping with malt syrup bait traps. Numbers of males and females are about equal.

a bethylid; and *Pardiaulomella ibseni* Gir., an eulophid. *Nemorilla pyste,* collected in northern Mexico, was released, and the predaceous carabid beetle, *Callida decora,* was also tested.

General predators exert some influence on GLF populations. Many species of spiders feed on larvae, while other species capture adults in webs. Wasps (*Polistes* sp.) collect GLF larvae as food for their larvae, but these wasps are not numerous under ordinary vineyard conditions. Birds, primarily blackbirds, have been observed feeding heavily on larvae only after vineyards were essentially defoliated, so their activity came too late to be beneficial.

Monitoring Guidelines

Monitoring guidelines have not been developed. With experience, however, there has been developed an intuitive sense about the incipient population based on the numbers of young larvae in samplings. However, this perception cannot be defined nor can the places exactly where to look be given and the appraisal is not always correct. Some years broods become worse than anticipated and in other years they are not as severe.

Growers are sometimes caught unaware of a damaging population and complain their vineyards were rolled up "overnight." While a great many rolls may be made in a short time, the incipient population is evident from seven to ten days before leaf rolling. The moths themselves show some flight activity during daytime or can be seen at night around porchlights. The eggs may be hard to find, but the larvae are not; their characteristic group feeding can easily be recognized at least a week before leaf rolling. Even if these signs are missed, some leaf rolls begin to appear before the bulk of the brood shows leaf rolling activity. These first rolls appear on the upper portion of the vine where they are easily seen, and vines should then be examined for small larvae.

Grape leaffolder populations are not uniformly distributed in a vineyard. One area may harbor a large infestation, while another may show a much lower infestation. But the same areas in a vineyard generally tend to be infested year after year and can be watched as guides to population trends. GLF and other pests of grapes, with the notable exception of Pacific spider mite, prefer vigorous, strong, densely foliaged vines.

Some vineyard managers may wish to trap GLF adults as a guide to populations expected or to follow the broods. The moths are easy to catch in bait traps containing fermenting syrup or terpinyl acetate, or with blacklight traps. It is questionable whether the time spent in trapping is worthwhile, except as a research tool. Trapping requires a lot of time. No one has yet shown that the traps give a useful guide to potential populations nor can the flights be interpreted except in retrospect.

Studies have shown that virgin GLF females attract males by releasing a sex pheromone. Virgin females can be used to attract males which are then captured on a sticky surface, but obtaining supplies of virgin females is inconvenient for any practical purpose. There is no synthetic sex pheromone available for trapping males.

Management Guidelines

Judging about treatment. In the past growers were encouraged to treat first brood populations on the premise that this was the single most effective treatment possible. This is now considered invalid; there does not seem to be a correlation between the past season's population and the current season's first brood nor with the population density that may develop later. Neither does treatment of the first brood preclude the need for treatment of later broods. In general each brood increases one- to fivefold over the previous brood, but there are many instances in which the population decreases. Thus, each brood must be inspected and judged as to its potential seriousness.

Treatment guidelines have not been developed. If population levels appear to be increasing, the vineyard should be examined every two or three days while larvae are making leaf rolls. If the brood becomes worse than expected, treat.

Typical populations are unevenly distributed. Decisions must be made whether or not to treat for a few bad spots. In wine or raisin grapes a few defoliated vines represent little loss. Table grape growers with a similar population would probably treat because the fruit is more susceptible to damage from exposure and sunburn.

Timing applications. First brood: Treatment rarely is required, but if necessary apply during bloom period, ordinarily May 10-25.

Second and third broods: Apply just as bulk of brood begins to make new leaf rolls. Remember that there may be a few off-brood larvae. To determine whether the leaf rolls are being made by the first larvae of the main brood, look for eggs and at the numbers and sizes of the larvae feeding in old leaf rolls or between leaves. If almost all eggs have hatched and if many larvae are third stage, treat.

The necessity for careful timing depends upon the insecticide used. *Bacillus thuringiensis* preparations such as Dipel, Thuricide, and Biotrol XK perform best when applied three or four days before the beginning of leaf rolling activity of the main brood, since their residual activity is fairly short. Carbaryl* (Sevin) will perform well if timed as above or if applied a week or so earlier because it has longer residual activity. Both materials will give fairly good emergency control even after many larvae are in leaf rolls, should this prove necessary.

In typical years second brood treatments fall in the period July 5-15 and third brood treatment in the period from August 25 into early September. Times vary according to the year and the particular vineyard involved.

If emergency treatment is required because too many rolls are appearing or if larva are entering the fruit, treat immediately. Good application is especially important if the insecticides must penetrate the leaf rolls or the fruit.

How to treat. For second and third brood it is important that the insecticide be directed as much as possible to the tops of the vines because this is where the leaf rolls are made. Growers who use insecticide dusts have altered their equipment so that the discharge comes down on top of the vines rather than up from underneath the vine canopy. Because coverage is vital, each side of the row should be treated.

Aerial applications are satisfactory for GLF because the insecticide is deposited on the top of the vines where the larvae concentrate. Late in the brood, when most of the larvae are in leaf rolls, aerial

application may not result in as good a deposit inside the leaf rolls as with good ground application, but it is certainly adequate for emergency treatment.

Amount of control obtained. Many observations and field trials show that a 90 to 95 percent population reduction can be attained in a moderately damaging infestation. With heavy populations, reductions range from 80 to 90 percent. Some larvae always survive a treatment; the heavier the infestation, the larger the numbers that get through.

Effects of some insecticides on GLF population. Some observations on the behavior of pest populations following use of insecticides may be useful. These observations are concerned with the insecticides, parathion* and carbaryl (Sevin), and B.T. (*Bacillus thuringiensis*) preparations, Dipel, Thuricide and Biotrol XK.

Parathion, extensively employed for GLF control before carbaryl became available, gave good reductions if properly timed, but any succeeding brood showed an unexplained population increase that again required control. Neither carbaryl nor B.T. shows this effect. However, retreatment of consecutive broods may be required.

Carbaryl, viewed solely for GLF control, is currently the most versatile material. However, in some vineyard districts it may increase Pacific spider mite populations. Carbaryl may also contribute to such other biological disruptions as mealybug outbreaks. B.T. preparations have been extensively tested for GLF control as a possible replacement for carbaryl. The currently available B.T. preparations are as effective as carbaryl, if properly timed. Thus, where actual or potential Pacific spider mite problems exist, B.T. should be applied in place of carbaryl. B.T. does not reduce populations of beneficial mites predatory on Pacific spider mite nor does it increase the reproductive potential of Pacific spider mite as sometimes may occur with carbaryl applications.

---

*Restricted material; permit required from County Agricultural Commissioner for possession or use.

Orange tortrix moths and eggs; female is right and male is left.

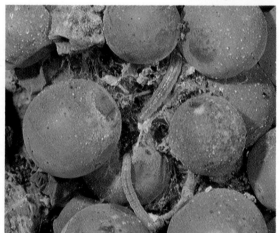

Orange tortrix larva in bunch.

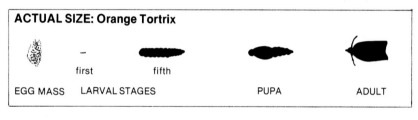

ACTUAL SIZE: Orange Tortrix

EGG MASS	LARVAL STAGES		PUPA	ADULT
	first	fifth		

Closeup of orange tortrix larva.

Mustard cover crop is one of the hosts for orange tortrix. Inset shows closeup of orange tortrix larval nest on mustard leaf.

The orange tortrix, *Argyrotaenia citrana* (Fernald), has been known in California since 1889, when it was found infesting orange trees, willow and goldenrod in Los Angeles County. Since then it has been noted on many different plants, but only since 1968 has it been seen as a pest on grapes. It is also a pest on citrus and deciduous trees in many areas and, because of its presence on apples in northern California, it has been known as "apple skin worm."

Geographical distribution of the orange tortrix ranges from Humboldt to San Diego counties throughout the coastal areas and in the interior valleys of the coast ranges.

Spread of this pest on grapes was first noted in 1968, when larvae feeding on grape clusters were found causing considerable damage in Salinas Valley vineyards in Monterey County. Since then this insect has been found on grapes in San Benito, Santa Clara, San Luis Obispo, Santa Barbara, Sonoma and Napa counties.

Establishment of the orange tortrix in coastal vineyards may be attributed to several factors: (1) The insect was present in general areas where grape plantings occurred; (2) it is able to feed and develop on a wide variety of plants and readily adapts to grape; and (3) the introduction and continuous use of herbicides in place of traditional row plowing to control weeds may have contributed to a buildup in grapes. Larval populations appear to survive and develop in numbers in under-vine trash that used to be destroyed by row plowing.

Description

The orange tortrix moth is variable in size, color and wing markings. When the moth is at rest, its wings are bell-shaped. The female moth is approximately 13 mm (1/2 inch) long and orange-brown; generally, she exhibits V-shaped or saddle-shaped darker colored markings in the middle of the wings when at rest. The male moth, slightly smaller than the female, is light brown or orange-brown; he also has a distinct V-shaped marking on the center of each fore wing, with a pair of crescent-shaped markings on the wing margins. The hind wings are white.

Eggs are flat, oval and cream colored and are deposited in masses, overlapping like shingles or fish scales. The number of eggs in a cluster may vary from a few to more than 200. On grapevines the eggs

are laid on any smooth surfaces such as upper leaf surfaces, stems or canes, or on the berries.

When first hatched, the larva is about 2 mm (1/12 inch) long, and fully grown it is approximately 13 mm (1/2 inch) long. The larva is generally straw colored or may be greenish or dark gray. The head capsule and the prothoracic shield adjacent to the head remain brown throughout the various larval developmental stages. The larva webs a nest among the plant parts in which it lives. When disturbed, it may wiggle violently and then drop suspended on a silken thread.

The pupa is about 13 mm (1/2 inch) long. Pupation generally takes place in the area where the larva has been feeding. It ceases feeding, constructs a silken shelter and becomes quiescent. It then contracts slightly and the last larval instar skin is molted. The pupal case is initially cream colored but within a few hours it turns dark brown. Females may be differentiated from males as being slightly longer and more robust.

Injury

The prime concern in an orange tortrix infestation is the development of grape cluster rot caused by fungi, molds and bacteria gaining entry into the fruit through larval feeding injury. Other types of damage may also occur. In early spring the overwintering larvae may attack emerging grape buds or young shoots. The damage may be confused with that caused by cutworm. Generally the presence of webbing points to the orange tortrix larva as the cutworm does not spin a web.

The larva often webs small leaves together to form protective nests in which it lives and feeds on the foliage. When fruit clusters form, the larva moves to them. Small clusters or portions of clusters may be webbed together. In the clusters the larva may feed on individual berries or stems so that cluster portions below the injury are killed.

In heavily infested vineyards during the dormant period the larva may be found feeding under the bark of canes that have twisted and split during cane tying.

Infestation appears more severe in varieties or vines with compact clusters. Pinot blanc, Chardonnay, Gewürztraminer and Pinot noir grapevines have been more frequently attacked than have other varieties in the central coast counties.

Seasonal Development

All developmental stages of the orange tortrix are present throughout the year. The overlapping generations make it difficult to estimate the size of any one generation. Investigations in central coast areas have shown there are three generations during the year: from late April to early May, in July and from October to November.

The orange tortrix is adapted to the cooler coastal regions. Laboratory observations showed that the minimal developmental temperature for the insect was 6°C (43°F), and the optimal, 25.6°C (78°F). Larvae failed to develop at temperatures above 32.2°C (90°F).

In spring larval infestation occurs in the developing foliage in the crowns of vines. Small larvae are found along the veins at the leaf base, and large larvae are found nesting in webbed leaves. In mid-July the infestation shifts from the vine centers to lateral areas. When the flowers and fruit clusters begin to form, larvae leave the foliage and invade the clusters; after harvest they are found in the surrounding vegetation and in clusters left on vines.

The orange tortrix does not hibernate (diapause) during winter. Most overwinter in the larval stages and are found in mummified clusters on vines or on the ground. Larvae are often found in such weeds as mallow, curly dock, filaree, lupine, mustard and California poppy; oats and barley grown as vineyard cover crops are also attacked by larvae.

The population gradually declines in winter because of adverse weather, insect predators and parasites, and birds feeding on the fruit clusters. The population begins to increase in early spring, reaching a peak in fall.

Natural Control

Examination of larvae and pupae collected from vineyards in the Salinas Valley show an ichneumonid wasp, *Exochus nigripalpus subobscurus* Tow., as the dominant parasite, constituting 95 percent of the parasites found. Three other parasites also found are a braconid wasp (2 percent), *Apanteles aristoteliae* Walker; a tachinid fly (2 percent), *Nemorilla pyste*

Walker; and a chalcid wasp (1 percent), *Dibrachys cavus* Walker. All are internal parasites of the orange tortrix larva. Parasitized larvae do not die until pupation and therefore they continue feeding on and damaging grapevines. Monitoring in 1980 in the Soledad area showed that moderate to heavy levels of parasitization in late spring were followed by low tortrix levels until harvest.

The *Exochus* wasp, a slender wasp about 7 mm (1/4 inch) long, has a black head and body and yellow legs. The female wasp parasitizes the orange tortrix larva by inserting an egg into the host's body cavity. The wasp larva then develops and pupates within the orange tortrix.

Because the moth and wasp emerge differently, the amount of parasitism in the vineyard can be estimated by examining the orange tortrix pupal case. The wasp emergence hole is circular, while the moth emerges through slits created on the sides of the pupal case.

The *Apanteles* wasp is about 4 mm (1/6 inch) long with a black head and body and yellow legs. It also parasitizes the orange tortrix larva by inserting an egg into the body cavity. The wasp develops within and kills its host when it leaves to pupate. It spins a white cocoon and pupates in it.

The *Nemorilla* fly not only parasitizes the orange tortrix larva but also the omnivorous leafroller and the grape leaffolder larvae. This fly, about the size of the common housefly (5 to 8 mm), is distinguished by a gray body with five black stripes on the thorax. The abdomen's sides are reddish. The female fly deposits one to several pearly white eggs around the head and thoracic regions of the orange tortrix larva. Upon hatching, the fly larva invades and develops within the host's body. The fly then pupates in its own puparium inside the pupal case of the orange tortrix.

The chalcid wasp, *Dibrachys* sp., is a small, dark colored wasp about 2 mm (1/12 inch) long. A number of these parasitic wasps may develop in and emerge from a single host. The *Dibrachys* wasp attacks many insects and is known to be hyperparasitic; that is, it also parasitizes other wasps.

The total effect of predatory insects and spiders in reducing orange tortrix infestations in the vineyards is not known. Some insect predators are: the damsel bug, *Nabis* spp.; the minute pirate bug, *Orius*

Orange tortrix larva feeding on developing bud. Note webbing.

Orange tortrix nest (arrow) in ripening bunch.

Orange tortrix feeding damage to berries and stem of cluster.

tristicolor (White); the big eyed bug, *Geocoris*, spp.; and the green lacewing, *Chrysopa* spp.

Monitoring Guidelines

Guidelines for treatment levels of infestation have not been established for the orange tortrix; however, some methods for examining vineyards to detect infestation are recommended.

First check vineyard areas where infestation is suspected or is known to have occurred. Varieties or vines having compact clusters are favored by the larva—Pinot blanc, Chardonnay, Gewürztraminer and Pinot noir are frequently attacked and should be examined. Webbing on the plant is a good indicator of larval infestation and should be examined more closely for evidence of orange tortrix. In spring larvae infest developing shoots and later they prefer flowers and fruit clusters. In infested clusters berries may show signs of wilting because of stem injury, or feeding injury may be seen on individual berries. Rotted portions of clusters should also be examined because orange tortrix may have initiated the rot.

In the dormant period weeds in the vineyard should be examined as well as clusters remaining on vines and on the ground. Such weeds as curly dock, filaree, lupine, mallow, California poppy and mustard, as well as oats and barley grown as cover crops, are likely host plants.

Commercial insect traps and lures are available, but their usefulness is limited because the relationship between moth catches and vineyard infestation is not known.

Management Guidelines

If orange tortrix is found in the vineyard, follow these procedures to reduce or prevent serious infestation:

Winter cleanup. Cleanup of the vineyard during the dormant period should be the initial control measure. Because most of the overwintering insect population will be in the larval stage on weeds or in dried grape clusters on vines or on the ground, it is important to remove these clusters when pruning. Place them in the middle of the rows where discing will bury them. Remove weeds and trash by row plowing and then by discing the vine rows. Do this work at least a month before shoots begin to develop in spring.

Early harvest. Observations have shown that larval populations increase in late summer and fall. In infested vineyards, therefore, it is important to harvest the crop as early as possible.

Insecticide treatments. Several insecticides are registered for control of orange tortrix on grapes. If infestation is not widespread, treatments may be confined to infested areas. Thorough insecticide coverage of vines is necessary, and each side of the row should be treated.

In more widely infested vineyards, control of orange tortrix should include vineyard sanitation as outlined and then an early season insecticide application sometime from March to May. Additional treatments may be necessary later in the season.

The grape mealybug, *Pseudococcus maritimus* (Ehrhorn), was a pest (primarily of table grapes) for many years in California, but it is not widespread today. Early control in the 1920s consisted of fumigating vines with sulfur under tarps. Before the 1940s occasional losses occurred in table grapes; even though bothersome, infestations were mostly spotty and frequently disappeared the following year. Increasing and more persistent grape mealybug populations were noted in the late 1940s, starting in the southern San Joaquin Valley's Delano/Earlimart table grape district and then in other grape areas. Extensive use of DDT and other new synthetic pesticides to control grape leafhoppers apparently had upset the grape mealybug's natural enemies.

Although mealybug infestations increased with general use of synthetic pesticides, the organic phosphate insecticides used after grape leafhopper populations became resistant to DDT seemed less disruptive. In fact, grape mealybug populations noticeably subsided in the late 1950s and early 1960s as DDT was phased out. Individual vineyards still suffered losses to mealybugs but less severely, and in many cases treatments were reduced or eliminated. Mealybugs remained a problem in table grapes, where heavy treatments were applied to maintain fruit clean of leafhopper spotting, but today they are a minor problem industrywide, only causing severe losses to an individual table grape grower.

Grape mealybug, a widely distributed species, is found on other crops, including walnuts, pears and pomegranates. Its greatest importance in California is as a pest of grapes, particularly in Kern County's Arvin district, Kern and Tulare counties' Delano areas, and Tulare County's Exeter vicinity. It is sometimes found in the Salinas and Santa Clara valleys, but it is not considered important there. Mealybug outbreaks have also been reported in Napa Valley vineyards in 1978 and 1979.

Description

Grape mealybug has a soft, oval, flattened, distinctly segmented body, although divisions among the head, thorax and abdomen are not distinct. The adult female, about 5.0 mm (3/16 inch) long, appears smoothly dusted with a white, mealy wax secretion. She has rather long caudal filaments along the lateral margin of the body that become progressively shorter toward the head. As many as 600 eggs are laid in a loose, cottony mass called the ovisac. The oval eggs are yellow to orange and can be seen with the naked eye. Crawlers that hatch from them range from yellow to brown and they are free of the waxy coating characteristic of later growth stages.

Females and males are similar in early instar stages. All stages of the female appear similar. The male passes through three nymphal instars, then forms a flimsy cottony cocoon about 3 mm (1/10 inch) long in which is formed the pupal stage, then the adult. The adult male has a pair of wings, a pair of halters provided with hooks, and two long, white, anal filaments. With the exception of the eggs and male pupa, all stages of mealybugs are motile.

Injury

The grape mealybug does no known harm to the plant, but it contaminates fruit with one or more of the following: the cottony ovisac, eggs, immature larvae, adults, honeydew or black sooty mold growing on honeydew. (The presence of honeydew is usually, but not always, evidence of grape mealybug damage. Some scale insects that occasionally infest vines also produce honeydew. Among honeydew-producing scale insects are European fruit lecanium scale, the most common, and cottony cushion scale, brown soft scale, cottony maple scale, frosted scale and black scale.) Level of damage is related to amount of contamination as well as to the use to be made of the

---

ACTUAL SIZE: Grape Mealybug

EGG   ADULT-FEMALE

---

fruit. In severe cases fruit will be unsightly, difficult or impossible to process and will have to be rejected.

Mealybugs are by far most serious and persistent in table grape vineyards. Their presence and honeydew cause fruit to be culled. Table grape vineyards are usually more likely to be infested because the pesticides used to insure clean fruit sometimes interfere with natural controls. Raisin or wine grape vineyards may be heavily infested at times, but the populations usually decline as rapidly as they explode. Also, they can tolerate higher populations; raisin processing removes some mealybugs and honeydew (if infestation is not too heavy) and wine grapes with fairly high levels of infestation may be diverted to distilling material.

Susceptibilities to mealybugs vary by variety, depending upon vine growth and fruiting characteristics, season of maturity and type of pruning involved. Vigorous vines are more likely to be infested than weak ones. As far as fruit is concerned, varieties such as Ribier that produce clusters close to the base of the shoot with fruit touching the old wood (arms or cordons) are likely to have more heavily infested clusters than are varieties where the clusters hang freely.

Early harvested varieties (Perlette or Cardinal) are much less likely to have serious fruit damage than late maturing varieties (Calmeria and Emperor). Early grapes are harvested before or just about the time the summer brood hatches and thus escape infestation. Late-maturing varieties remain on the vine long enough for the summer brood to mature and are more likely to contain the cottony ovisacs at harvest, along with honeydew, sooty mold and the mealybugs themselves.

Cane-pruned varieties (Thompson Seedless) are less likely to be seriously infested than are spur-pruned vines. Because mealybugs overwinter on old wood, the young have to move further to disperse over the vine. Cane-pruning protects the fruit a little more in relation to its position to old wood where most mealybug eggs are laid. For the summer brood, however, some mealybugs lay their eggs on fruiting canes rather than on the old wood of the head or trunk.

Seasonal Development

The grape mealybug overwinters on old wood under loose bark in vineyards either as eggs or as newly hatched crawlers in or near the white, cottony ovisac. In spring most crawlers move toward the base of the spurs and then onto the expanding green shoots, reaching maturity in mid-May to early June. Most females return to old wood to lay the eggs that hatch from mid-June to early July. The first generation crawlers then move out to the green portions of the vine to feed on fruit and foliage. It is this brood that produces almost all fruit damage. Some females maturing in late August and September lay their eggs on fruit and leaves; however, most return to old wood to lay overwintering eggs.

There are two broods a year. While the bulk of the two broods makes a round trip from the old wood to the leaves and fruit and back again, a small number never seem to leave the protection of old wood. Somehow they manage to survive, possibly by feeding at the bases of spurs, on callus tissue at the sites of girdles, and below the old bark itself. They apparently reach live tissue, however, because fresh honeydew can be observed.

The overwintering brood that goes out on the foliage in spring is usually present in low numbers. The summer brood, hatching in June, can show a tremendous population explosion. Serious damage may result when little evidence of infestation was present in the spring.

Natural Control

Little work has been done in grapes to determine the effectiveness of natural enemies of grape mealybug. They must be responsible for keeping populations at low levels because a few mealybugs can be found in almost all vineyards, but only a small percentage ever requires control. The following parasitic wasps have been identified as attacking grape mealybug in the San Joaquin Valley: *Acerophagus notativentris* (Girault), *Anagyrus yuccae* (Coq.), *Zarhopalus corvinus* (Girault) and *Pseudleptomastix squammulata* (Girault). The effectiveness of these endoparasitic wasps of the family Encyrtidae seems to vary considerably in place and time, according to limited data and observations. At times any one of them may exert considerable parasitism on mealybug populations, but little is known about multiple or long range parasitism, particularly as influenced by the use of pesticides for other pests.

Less is known about the effectiveness of predators of grape mealybugs in vineyards. In 1979 a Cecidomyiid

Grape mealybug egg mass under bark. Inset shows grape mealybug female.

Infested Emperor bunches touching bark.

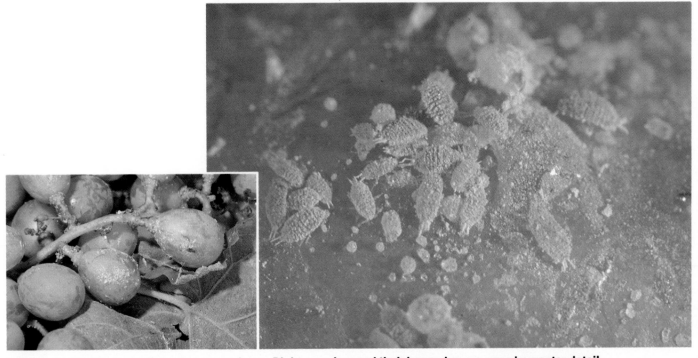

Left, grape mealybug crawlers are seen on berry. Right, crawlers and their honeydew are seen in greater detail.

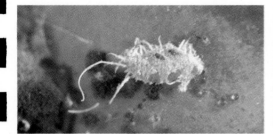

Emergence holes of a grape mealybug parasite.

Cryptolaemus ladybird beetle larva feeding on grape mealybug egg mass.

Comparison of grape mealybug female, left, and Cryptolaemus larva, right.

fly larva in Tulare County and *Cryptolaemus montrouzieri* Mulsant (the mealybug destroyer) in Napa Valley were observed attacking grape mealybug eggs. Collection data indicate that the fly larva may play an important role late in the season. Green lacewing adults, frequently abundant on grapevines harboring mealybugs, are attracted to mealybug honeydew, but to what degree their egg laying and subsequent control of mealybugs is influenced by the presence of honeydew is not known. Other predators associated with grape mealybugs are sympherobiids (brown lacewings), *Orius* (minute pirate bug), predaceous beetles, nabids and several species of spiders.

Monitoring Guidelines

Like most grape pests, with the notable exception of Pacific spider mites, grape mealybugs prefer vigorous vines. Thus, the most likely infestations are in vigorous rows and end vines. Weaker vines may harbor mealybugs but usually populations will not be heavy. While serious infestations are largely confined to the more vigorous vines, some vigorous areas of a vineyard may be infested and others not.

Spring monitoring. From April to June almost all populations of any consequence can be detected fairly easily by the presence of honeydew and/or the black sooty mold on leaves. Very light populations may require microscopic examination of leaves.

Summer monitoring. By June the white ovisacs containing eggs or newly hatched young can be found under old rough bark near the spurs. Then as the young begin moving out, they may be found on clusters touching old bark and on leaves or spurs close to the bark. Later, when these young mealybugs disperse, they are most easily detected by their fresh honeydew deposits that are clear, sticky, glistening, small droplets. As the mealybugs grow, the honeydew droplets become larger and begin to develop the characteristic sooty mold. Sometimes growers do not realize the presence of the summer brood mealybugs until the fruit is unsightly; early detection is important. Examination (while monitoring leafhoppers) should be made during July 1-15 of clusters touching old wood on vigorous vines. If no young mealybugs are seen on these clusters, little or no infestation is present. If young mealybugs are found, examine other vigorous areas of the vineyard, especially those that have been infested in the past.

Preharvest monitoring. Monitoring for dormant treatment decisions is done just before harvest. The procedure in vineyards more than 40 acres consists of inspecting for the presence of mealybugs on five bunches on every tenth vine in every thirtieth row. (A proportional sample size is taken in vineyards of less acreage.) Special attention is given to those bunches contacting the bark. Bunches are recorded as infested, if any signs of mealybugs are detected, even if they would not be culled.

Winter monitoring. The undersides of the arms on spur-pruned vines are favorite overwintering areas; fewer mealybugs will be found on the cordon or trunk. With cane-pruned Thompson Seedless the greatest number of mealybugs are on the head and upper trunk area, but they do disperse down the trunk to the soil. On trunk-girdled vines there are usually mealybugs around the girdled area feeding on the callus tissue. During winter vines that were infested the previous season may appear noticeably darker than uninfested vines because of the presence of black, sooty mold.

Management Guidelines

The decision to treat grape mealybug is difficult to make and must depend on knowledge about previous mealybug problems in a given vineyard. Close monitoring through the year is essential, at least in problem areas, because an infestation can develop rapidly with little warning or a heavy infestation may decline dramatically as a result of natural controls.

With the materials available and the dosages permitted the best single control is usually obtained with a dormant spray. However, if infestation is high, it may be necessary to follow with a summer treatment. Spring treatments have proved ineffective.

Midsummer treatments used alone give substantial protection for the fruit but cannot be expected to reduce overwintering populations. Thus, summer treatment should ordinarily be followed by a dormant spray. Summer monitoring procedures should be followed closely to determine whether mealybugs pose a threat. No treatment levels have been developed. The grower or his advisor will need to use his best judgment. If the summer brood needs to be treated, controls must be applied when the mealybugs are small and vulnerable so as to kill a high proportion. The longer treatment is delayed, the less effective control will be. In fact, once mealybugs are half grown, it is believed that controls are not worth applying.

When fruit losses at harvest appear to be approaching the cost of treatment or when preharvest monitoring shows more than 2 percent of the bunches with any sign of mealybugs, dormant treatment is usually warranted. However, the vineyard should be examined again after harvest to determine the need for a dormant control. Sometimes heavy infestations in September and October are eliminated by natural controls. Someone able to recognize live eggs or young mealybugs should check the vines because the mere presence of cottony ovisacs—which can persist for a year or two—does not automatically mean that live mealybugs are present.

Finally, each grower or advisor must make his own decisions. The more information one has about the general behavior of mealybugs and their natural enemies and of behavior in a particular vineyard, the better the decisions made.

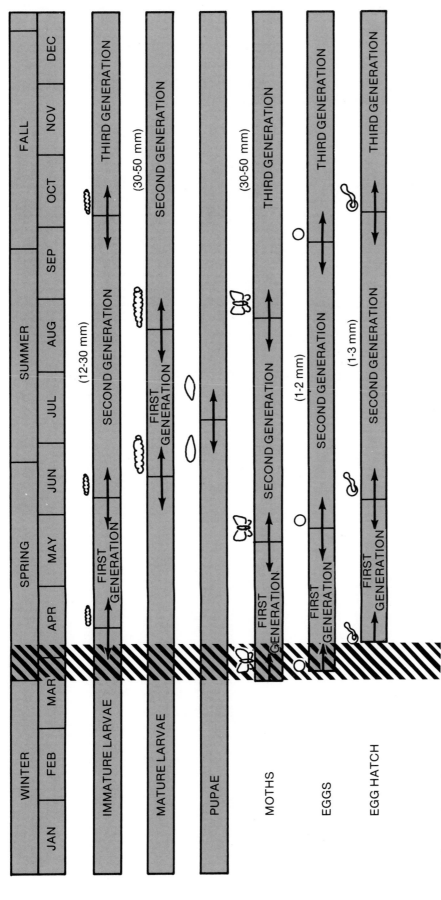

CUTWORM LIFE CYCLE

Major cutworms in grape pest management:
Greasy cutworm—northern states, two generations; southern states, four generations. Overwinters in larval and pupal stages.
Brassy cutworm—one generation. Overwinters as mature larvae.
Variegated cutworm—three to four generations. Overwinters in larval and pupal stages.
Spotted cutworm—two to three generations. Overwinters in larval stage.

Key:

= Most important period for grapevine damage (bud and early shoot growth stage).

= Overlapping generations.

Aerial view of Cabernet Sauvignon vineyard showing phylloxera damage (arrow).

Ground level photo of same spot on aerial photo.

Table Thompson Seedless production. Thompson Seedless table grapes require special cultural practices to produce commercially acceptable fruit. The trellis always includes one crossarm and sometimes a double crossarm. This spreads the fruit and makes it accessible for thinning and harvesting.

The fruit is sprayed with gibberellin at a concentration of 10 to 20 ppm and at a rate of 150 gallons per acre in the 30 to 80 percent bloom stage. This is called a bloom thinning spray and its primary purpose is to reduce heavy set and tightness of clusters; it also elongates and enlarges the berry a little.

After fruit set is completed vines are girdled and resprayed with gibberellin. This treatment is called the berry-sizing spray. A 20 to 40 ppm spray is applied at a rate of about 200 gallons per acre. Usually two such sprays are applied, the second five to seven days after the first. The fruit is also thinned by removing entire clusters or by removing cluster parts. The most common method of berry thinning is to remove all but the basal four to six laterals.

Natural Thompson (those not girdled, treated with gibberellin or thinned) usually weigh 1.4 to 1.8 grams per berry. Table Thompson Seedless range from 3.5 to 5.5 grams per berry, depending upon treatment used.

Description

Species of thrips considered to be grape pests are: The western flower thrips, *Frankliniella occidentalis* (Pergande); the grape thrips, *Drepanothrips reuteri* Uzel; the citrus thrips, *Scirtothrips citri* (Moulton); the bean thrips *Caliothrips fasciatus* (Pergande); and a minute flower thrips, *F. minuta* (Moulton). A predaceous thrips is also found on grapes, the sixspotted predaceous thrips, *Scolothrips sexmaculatus* (Pergande). It is an important predator of Pacific spider mites; western flower thrips has also been observed feeding on Pacific spider mite eggs.

Thrips commonly found on grapes can be identitied by using Table I and Figure 2. They are about 0.8 to 1 mm (1/25 inch) long.

Injury

Western flower thrips damage on grapevines consists of: (1) halo spotting that can make the fruit of certain white varieties unsightly and unmarketable; (2) ber-ry scarring on Thompson Seedless that also can render them unsaleable; and (3) shoot stunting and foliage damage. Grape thrips is mainly responsible for summer foliage damage, although it occasionally causes troublesome fruit scarring and shoot stunting. White Malaga grapes are particularly susceptible to fruit scarring by grape thrips.

Fruit Damage by Western Flower Thrips

Halo spots. Halo spots are the result of ovipositing (insertion of eggs into the tissue of small berries). A small dark scar is produced at the site of the puncture. The tissue in a roughly circular area around this puncture becomes whitish. As the grapes grow, these spots may show various growth cracks on the large berried varieties.

If only a few halo spots are evident, as is the case with most table grape varieties, no cullage results. However, some varieties are prone to halo spotting. These are the "white" varieties: Almeria, Calmeria and Italia. Of these, halo spotting is most serious with Italias because the skin of the berries at the spot may crack during ripening, allowing rot to develop. On any other variety rot does not normally result from thrips scars.

Scars without halo. On varieties subject to severe halo spotting, many small dark scars are sometimes seen without the surrounding halo. This scar's source is not definitely known, but it is believed associated with the thrips as scars do not occur when western flower thrips populations are controlled. It could be the result of probing by the female without egg deposition, the deposition of an infertile egg, or an ovipuncture made near the end of egg laying when the response of berry tissues may be different. The first suggestion seems most reasonable. A similar scar has been produced by puncturing berries with a fine wire.

With a few halo spots present the number of dark scars is usually well below the numbers of halo spots. With numerous halo spots there may be an equal number of dark scars.

Time when halo spots are produced. Studies of Italia show that halo spotting is produced both in bloom and up to fruit set or shortly thereafter. This period may last from ten days to two weeks or more, depending upon temperature.

TABLE I. Key to Thrips on Grapes.

If speciman shows: Proceed to step:

Step 1   Abdomen with dorsal-lateral microsetae (Figure 2A) 2

          Abdomen without dorsal-lateral microsetae 3

Step 2   If antennae are six-segmented, the thrips is:
          Drepanothrips reuteri Uzel (Figure 2A)
          If antennae are eight-segmented, the thrips is:
          Scirtothrips citri (Moulton) (Figure 2B)

Step 3   If body is reticulated, the thrips is:
          Caliothrips fasciatus (Pergande) (Figure 2C).
          If body is not reticulated 4

Step 4   If pronotum has prominent setae at anterior corners, midlateral setae are
          long and fore wings have six dark spots, the thrips is:
          Scolothrips sexmaculatus (Pergande) (Figure 2D).
          If pronotum has prominent setae at anterior corners, but midlateral setae are
          short and fore wings are without spots.. 5

Step 5   If interocellar setae are long, the thrips is:
          Frankliniella occidentalis (Pergande) Figure 2E)
          If interocellar setae are short, the thrips is:
          Frankliniella minuta (Moulton)

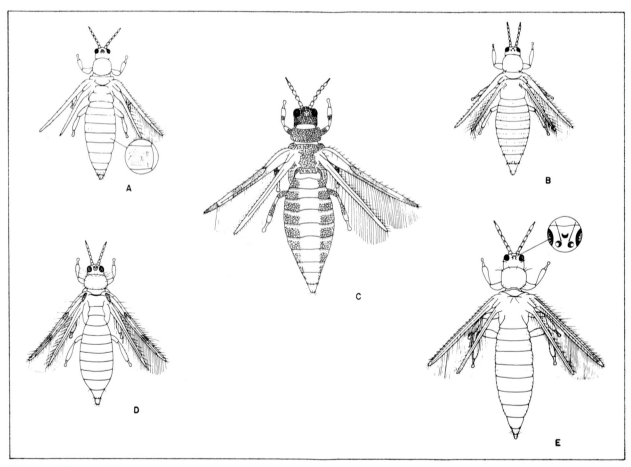

Figure 2. These thrips are reported to occur in grapes: (A) The grape thrips, *Drepanothrips reuteri* Uzel (insert shows detail of dorsal-lateral microsetae); (B) the citrus thrips, *Scirtothrips citri* (Moulton); (C) the bean thrips, *Caliothrips fasciatus* (Pergande); (D) a sixspotted thrips, *Scolothrips sexmaculatus* (Pergande); and (E) the western flower thrips, *Frankliniella occidentalis* (Pergande) with insert showing detail of relative length of interocellar setae.

Grape thrips injury to Barbera foliage.

Older grape thrips damage to White Malaga foliage.

Photo by Amand N. Kasimatis

Grape thrips damage to White Malaga berries.

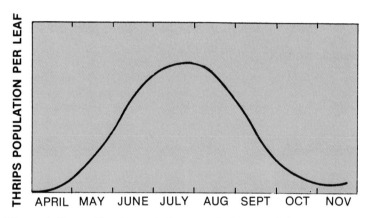

Figure 4. Generalized population trend of grape thrips on grape leaves. Peak populations are usually reached about mid-July. Peak and thrips numbers vary from year to year.

Vineyards with grass or other vegetation in or nearby usually support larger thrips populations (with the exception of grape thrips which is restricted to grapes as a host) and more likely to show foliar damage. Western flower thrips, for example, may move into vineyards from nearby areas when weeds are disced or may migrate from drying-up grasslands.

Damage in late spring and summer. Grape thrips, *D. reuteri*, usually do not produce much foliar damage in the early spring because their populations are too low. In May they occasionally damage Salvadors, a variety particularly vulnerable to them.

Generally growers are unaware of grape thrips until midsummer when the shoot tips may be attacked (Figure 4). Growth is stunted; leaves fail to grow to normal size and they are bronzed. The internodes are shortened and scarred. Usually the populations sub-

side after a brief peak and normal growth again resumes.

Adults and nymphs can be found in these damaged tips if they are examined when the damage is occurring. Often the population has decreased by the time the damage becomes overwhelmingly evident. Growers may be loath to believe the foliar symptoms were produced by the very few thrips that remain.

In most all cases the peak population of grape thrips comes too late in the season to warrant concern. Vines have developed a great enough leaf surface by mid-July to mature the crops and accumulate a good reserve for the next year. Only in a few cases is control warranted. Weak vines cease growth by early July; stronger vines that keep growing and produce new tip growth susceptible to grape thrips attack can tolerate damage although they do look peculiar.

Damage by Citrus Thrips

Citrus thrips, *Scirtothrips citri* (Moulton), have been reported as a pest of grapes only in Riverside County's Coachella Valley. Populations high enough to distort tip growth are most often encountered in late summer and fall. Their feeding is not usually considered damaging. Occasionally citrus thrips may scar young grape berries in the first few rows adjoining citrus groves. They seldom require control.

Seasonal Development

Western flower thrips. The western flower thrips is found throughout California on a great variety of plants. There are, reportedly, five to seven generations per year. Population peak is reached in May coinciding with grape bloom. This thrips overwinters in the adult and nymphal stages.

Reproduction may be either sexual or asexual. The minute eggs are laid in soft tissues of the plant, particularly in flowers. Each female lays about 20 eggs that hatch in about five days. The nymphs feed on the host through two larval (nymphal) stages lasting a total of seven to twelve days. Their prepseudopupal and pseudopupal stages last four to five days in the soil debris. Then, as adults, the thrips emerge and are attracted to grape blossoms, possibly by their fragrance and light color.

Adults feed on pollen, but it is not clear to what extent they feed on stem or fruit tissue. The nymphs feed only on stem tissue, if the flower parts have shed, or feed on both stem and fruit tissue if flower parts persist.

Usually only a few thrips are found on the leaves except shortly before and for about a month or so after bloom. They can, however, be found feeding on young shoots in the early spring, especially if the vineyard has weed or grass cover, or adjoins weedy areas or such crops as alfalfa.

There are three color forms of the adult female: light, intermediate and dark. The dark form is predominant in the early spring; the light and intermediate forms are most common later. The light form generally is the most numerous.

Males are numerous only in spring.

Grape thrips. Grape thrips originated in Europe, reportedly reaching North America in 1926. Only virgin females overwinter. They hibernate in the soil in low numbers, emerging when grapevines begin growth. Their first generation is produced asexually. By midsummer there are about eight females to one male; by fall the males disappear.

In California grape thrips has only been collected on grapevines and poison oak. In general, it prefers white grape varieties, especially tender foliage.

Eggs are laid in leaf and stem tissue. Larvae begin to appear in early April. By early May, adults begin to appear, most commonly on the upper surfaces of young leaves. Their population usually peaks in July but may vary from mid-June to mid-August. During this period the thrips concentrate on the shoot tips, especially on the tops of vines. Young leaves may be severely distorted and new growth restricted. Populations decline usually sometime after midsummer with only a few thrips remaining after early October. (See Figure 4.)

There are five to six generations each year with about 22 days required for a generation in midsummer. Average stage lengths are: egg, seven days; larvae (both nymphal stages), seven days; prepseudopupa, one day; pseudopupa, two days; preoviposition period, five days.

Like western flower thrips, the larvae drop to the soil to pupate after completing feeding. They locate in the debris under the vines or on the soil's surface.

Natural Control

Little is known about natural control of thrips in vineyards.

Monitoring Guidelines

Fruit. Western flower thrips populations are determined by counting adults or nymphs knocked out of the flowers or fruit clusters. This is done by sharply striking attached clusters three times against the flat surface of an 8-1/2x11-inch cardboard. During early cluster development adults are fairly well dislodged by this method but the nymphs are not. Once berry size reaches a diameter of 7.9 to 12.7 mm (5/16 to 1/2 inch), the fruit becomes difficult to manipulate and fewer thrips can be knocked out. Adults are easily and quickly counted, but nymphs are barely visible, especially at their first stage. To count them, the counting board is tipped to let the large flower parts fall off. Then the board is turned almost parallel to the sun's rays so that the nymphs cast shadows. It takes practice to distinguish slowly crawling nymphs from pollen grains and small flower parts. To make counting easier, use a counting board with a sectionalized or grid pattern. Also, nymphs are more easily seen on a light blue cardboard.

Peak counts will be in excess of 150 adults and 300 nymphs per cluster. More normal populations range per cluster from five to 25 adults and from 10 to 50 nymphs.

Specific monitoring guidelines have not been developed to interpret grape thrips population levels that are likely to scar fruit. Vineyards with histories of fruit damage by grape thrips bear watching carefully for thrips activity on young fruit. Either grape thrips or western flower thrips may be present on the young fruit. Identification is important on Thompson Seedless because western flower thrips will pose a problem only if there is a persistent cap problem as previously discussed. Particular attention should be given to examining clusters exposed to the sun; grape thrips seem to prefer them. Grape thrips also appear more active than the slower moving western flower thrips. Positive identification is made by using the key in Table I or by submitting a sample to an appropriate insect taxonomist. This service is provided by farm advisors and County Agricultural Commissioners.

Foliage. Early season western flower and grape thrips damage to the foliage and shoots reportedly occurs in the northern San Joaquin Valley and coastal valleys when vines leaf out early and when growth is slow during cool weather. It is not uncommon under such circumstances for vineyards to suffer many stunted shoots. Some vines may have half of their shoots stunted while others grow normally. Close inspection of stunted shoots will reveal high populations of thrips or their heavy scarring.

Because grape thrips is not usually considered troublesome during late spring and summer, monitoring guidelines have not been developed. However, as previously mentioned, a heavy grape thrips population may pose a problem in late spring on Salvadors and in weak vineyards during the summer.

Management Guidelines

Fruit injury. The amount of thrips damage allowed under U.S. No. I Standards for Table Grapes is not easy to specify. The Standards allow a total of 8 percent tolerance (by weight of damaged berries) for

several defects. Included are scarring by thrips, discoloration, heat injury, Almeria spot, mildew, or insect injury, infestation or residue. The Standards define damage as meaning "any defect which materially detracts from the appearance, or edible or marketing quality." Thus, there is a judgmental factor in interpretation of damage.

Italia: Most Italia variety vineyards require annual pesticide treatment for western flower thrips because halo spotting is frequently severe enough to lead to bunch rot. Unless past experience has shown little loss without control, treatments should begin at the 5 to 10 percent bloom stage and be repeated as necessary to keep adult populations below five per cluster until the fruit set stage. (See *Monitoring.*) Trials with the insecticide dimethoate (Cygon, Defend) require two treatments seven days apart during a year of normal bloom. In a warmer year with a fast bloom, only one treatment is required. In each case, the thrips population is greatly reduced for a week or slightly longer. Treatment frequency with other insecticides will depend upon performance.

Calmeria: Normally, less than one-fourth of the state acreage of this variety is treated each year. Halo spotting is not as severe as it is in Italias nor does rot normally develop from the halo spots. Many growers who have never treated suffer little loss; others are concerned because of high cullage. Based on limited experience, treatment is believed warranted only with a population of 10 adults or more per cluster in vineyards with histories of thrips injury.

When treatment is needed, insecticides should be applied at about 70 percent bloom. Most halo spotting occurs from this point until about fruit set. Trials with dimethoate show that one treatment is sufficient. Other insecticides may require more applications.

Almeria: Trials have not been run with Almerias, but when treatment is necessary, the guidelines for Calmeria should be followed. As Almeria berries do not grow as large as Calmeria, halo spotting is less serious.

Table Thompson Seedless: If caps stick in vineyards with a history of fruit scarring, treat when nymphs appear, at about 70 percent bloom stage. One treatment should be enough, but this may depend upon the chemical used. Severe damage can occur with peak populations of 10 adults per cluster and many persistent flower parts.

The performance of any material for reduction of halo spotting or berry scarring must be based on reducing damage as well as population control. Some insecticides may reduce western flower thrips population, but they may increase damage because thrips females seem to increase their ovipositing under stress of insecticide exposure.

Because grape thrips only occasionally poses a problem to table fruit, little if any testing on the timing of applications for controlling grape thrips has been done. However, both western flower thrips and grape thrips are controlled by the same chemicals.

Foliage injury. Western flower thrips and grape thrips in early spring and grape thrips during the summer are controlled by the same chemicals.

GLOSSARY
Common Insect and Mite Terms Commonly Used in Grape IPM

Abdomen—Posterior of three main body divisions of an insect.

Abiotic natural control—Control of pests by the action of nonliving things (for example, high temperatures).

Active ingredient—Chemical(s) in a prepared product responsible for effects; commonly used in abbreviated form, a.i.

Aedeagus—Male intromittent organ.

Aestivation—To rest, sleep or be inactive during hot summer months. An aestivating insect may be inactive for one or two months.

Anal—Pertaining to the last abdominal segment (which bears the anus).

Anal shield—Hard plate on terminal segment of a caterpillar and certain other immature insects.

Antenna—Pair of segmented appendages located on the head about the mouth parts and usually sensory in function.

Apical—At the end, tip or outermost part.

Apterous—Wingless.

Asexual—Reproducing without fertilization; parthenogenetic.

Basal—At the base; near point of attachment (of an appendage).

Biotic natural control—Control of pests by the action of living things (parasites, diseases, etc.).

Blacklight—Ultraviolet light that is visible to insects but not visible to man.

Brassy—Yellow with the luster of metallic brass.

Brood—All the individuals that hatch at about the same time, from eggs laid by one series of parents, and that normally mature at about the same time.

Cannibalistic—Feeding on other individuals of the same species.

Caterpillar—Larva of a butterfly or moth.

Chrysalis—Pupa of butterfly.

Cocoon—Silken case inside which pupa is formed.

Coxa—Basal segment of insect leg.

Crawler—Immature developmental stage in which the insect (scale or mealybug) has legs and ability to move, although in later stages it may become permanently affixed to one site.

Cultural control—Making the habitat unfavorable for the pest and/or favorable for natural enemies.

Degree-day—Unit of accumulated heat equal to 1 degree Farenheit above an average daily temperature (for grape leafhopper) of 50.5 degrees for an entire day (24 hours).

Deutonymph—Third instar of a mite.

Diapause—Dormant period between periods of activity.

Disruptive—Treatment that lessens the ability of beneficial species to exert controlling influences on a pest species.

Distal—Farthest from the body or main stem.

Dormant—That time of year when vines and most insect and mite species are inactive.

Dorsal—Top or uppermost; pertains to the back or upper side.

Economic injury level—Pest population level sufficient to cause economic losses greater than the cost of a treatment.

Economic threshold—Pest population level on the threshold of causing damage to a crop greater than the cost of a treatment.

Ectoparasite—Parasite that attaches itself to a host and feeds externally.

Elytra—Thickened, leathery, or horny front wing, characteristic of beetles.

Embryo—Young animal before leaving the body of the parent or before emerging from the egg.

Endoparasite—Parasite that feeds and develops within a host's body.

Entomophagous—Feeding upon insects; specifically applied to those wasps that feed their young with larvae, etc.

Erinea—Galls made up of tiny plant hairs with round heads. The gall develops in the form of a velvety pad. Caused by eriophyid mites.

Exoskeleton—External skeleton of insects and other arthropods. The exoskeleton is discarded (molted) at various times.

External parasite—See **Ectoparasite.**

Exuviae—Cast skin of an insect or mite. See **Exoskeleton.**

Femur—Third segment of the insect leg, located between trochanter and tibia.

Filament—Threadlike.

Flare-back—Resurgence of a pest species following use of a treatment that suppresses the effect of beneficial species control.

Flight—Period of adult moth flying activity; flight length may be selected to define a part or all of the flying activity of a generation of moths.

Flight peak—Point of maximum flight activity during a flight period.

Frass—Insect excrement.

Generation—From any given stage in life cycle to the same state in the offspring.

Granulosis virus—Virus particle contained within minute granules.

Gravid—Pregnant.

Gregarious—Living in groups.

Grub—Thick-bodied larva with well-developed head and thoracic legs, without abdominal prolegs, and usually sluggish.

Head—Anterior body region bearing the eyes, antennae and mouth parts.

Hibernation—Dormancy period. See **Aestivation.**

Honeydew—Sugary, syrupy substance secreted by aphids, mealybugs and soft scales.

Host—Organism in or on which a parasite lives; plant on which an insect feeds.

Hot spot—Vineyard area where a pest (mites) population is substantially larger than in balance of vineyard and is at or nearing the point of causing economic damage.

Hyperparasite—Parasite whose host is another parasite.

Indigenous—Native.

Instar—Period or stage between molts, the first instar being the stage between hatching and the first molt.

Integrated control—Control approach that uses all suitable techniques and methods (e.g., chemical, cultural and biological) in as compatible a manner as possible to maintain populations at levels below those causing economic injury.

Internal parasite—Endoparasite.

Larva—Immature stages between egg and pupa in insects; six-legged first instar of mites.

Life cycle—Series of changes in the form of life beginning with egg and ending in adult reproductive stage.

Mandibles—Stout toothlike lateral upper jaws of chewing insects; used for seizing and biting.

Metamorphosis—Change in form during development.

Millimeter—0.001 meter or about 1/25 inch.

Molt—With reference to insects and mites, the shedding of skin before entering another stage of growth.

Motile forms—Those forms or stages in the insect life cycle capable of movement.

Mottled—Marked with spots of different colors.

Nocturnal—Active at night.

Nodosities—Swelling or galls caused by grape phylloxera on the tips of rootlets.

Nymph—Stage of development in certain insects immediately after hatching, resembling the adult but lacking fully developed wings and sex organs.

Oblong—Longer than broad.

Omnivorous—Feeding on many hosts.

Opaque—Without any surface luster; not transparent.

Ovary—Egg producing organ of the female.

Ovigenesis—Formation and maturing of egg.

Oviposition—Act of laying eggs.

Ovipositor—Egg-laying apparatus.

Ovisac—Extension of the body wall or specialized structure of a female, where certain kinds of insects (mealybugs) deposit their eggs.

Papilla—A minute, soft projection.

Parasite—Animal or plant that lives in or on the body of another living animal (its host) at least during part of its life.

Parthenogenesis—Reproducing by eggs that develop without being fertilized.

Pest management—An approach to pest control in which populations are monitored and integrated control methods applied at a time that most effectively holds pests at manageable, prescribed levels.

Pheromone—A substance secreted externally by insects to affect the behavior or development of other members of the species; e.g., sex pheromone to attract and arouse members of the opposite sex.

Photophase—Day length.

Phytophagous—Feeding on plants.

Pinaculum—In caterpillars, an enlarged seta-bearing papilla forming a flat plate.

Polyhedrosis virus—Virus particles contained within minute polyhedric body.

Population-density criteria—Population levels selected for pest management sample counts to indicate probable relative abundance of population in the vineyard.

Predaceous—Feeding as a predator.

Predator—Animal that attacks and feeds on other animals (its prey). The prey is usually killed quickly and mostly or entirely eaten; many prey individuals are eaten by each predator.

Prepseudopupa stage—Third nymphal instar of thrips.

Prolegs—Appendages that serve as legs on abdomens of caterpillars and other fleshy larvae.

Pronotum—Upper or dorsal surface of the prothorax.

Prothoracic shield—Chitinous plate on prothorax of caterpillar just behind head.

Prothorax—Forward segment of an insect's thorax, bearing the first pair of legs.

Protonymph—Second instar of a mite.

Pseudopupa—Larva in a quiescent pupalike condition.

Pubescent—Downy; clothed with soft, short, fine, closely set hair.

Pupa—Stage between the larva and adult in insects, a nonfeeding and usually inactive stage.

Puparium—In flies, the thickened, hardened larval skin within which the pupa is formed.

Quiescent—Not active.

Random sample—Monitoring sample taken randomly in a vineyard.

Reduced rate—Lower dosage rate of pesticide than normally recommended but which works better in pest management programs because it allows greater survival of beneficials.

Resistance—Developed ability of insect or mite populations to withstand pesticide effects.

Reticulum—Any netlike structure.

Secondary pest—A pest that normally is not a problem until the effects of controls applied for another pest or pests enable it to increase and cause damage.

Setae—Slender hairlike appendages on insects and mites.

Sooty mold—A dark, often black, fungus growing in insect honeydew.

Spiracles—Lateral breathing pores on the segments of an insect body.

Standard rate—Dosage rate of a pesticide that is commonly recommended for control; usually selected for its effect on target species without consideration of impact on beneficial species.

Stipe—Stalklike part of Willamette mite egg.

Striated—Marked with a slight furrow, ridge or streak.

Sub-dorsal—Below the dorsum and above the stigmata (spiracle).

Tarsus—Leg segment beyond the insect tibia, consisting of one or more segments or subdivisions.

Taxonomist—Scientist especially trained in the classification of plants and animals.

Thylacium—External gall-like cyst in abdomen of host, containing parasitic larvae of Dryinidae wasps.

Tibia—Fourth segment of the insect leg, between the femur and tarsus.

Treatment level—Population level at any given stage of development or time of the year, when a treatment or other management practice will most effectively and economically prevent crop damage in either the current season or the subsequent crop year.

Translucent—Semi-transparent.

Trochanter—Second segment of the insect leg, between the coxa and the femur.

Tuberosities—Swellings caused by grape phylloxera feeding on larger roots.

Vein—Thickened line in wing.

Ventral—Lower or underneath; pertaining to the underside of the body.

Wing pads—Encased undeveloped wings of nymphs (leafhoppers) which appear as two lateral flattish structures behind the thorax.

(a) Beet armyworm larva.

(b) Western yellowstriped armyworm larva.

(c) Achemon sphinx moth larva, green phase.

(d) Achemon sphinx moth larva, brown phase. (Note loss of horn.)

(e) Saltmarsh caterpillar eggs, newly laid.

(f) Saltmarsh caterpillars, second and fifth instars.

Photo by Frederik L. Jensen

SECTION IV—MINOR INSECT AND MITE PESTS

Contents

Page

* Information on these minor pests was taken from Circular 566, *Insect Grape Pests of Northern California*, E.M. Stafford and R.L. Doutt, June 1974 (out of print).

+ Information on aphids supplied by Frederik L. Jensen and Hiroshi Kido.

WESTERN YELLOWSTRIPED ARMYWORM

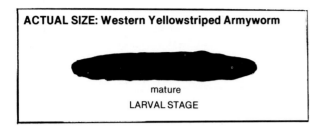

ACTUAL SIZE: Western Yellowstriped Armyworm

mature
LARVAL STAGE

The western yellowstriped armyworm, *Spodoptera praefica* (Grote), is primarily a pest of alfalfa. It also lays eggs and completes larval development on a number of broadleaf weeds. It is not known to lay eggs on grapes, but the larvae will attack grape foliage in vineyards adjacent to alfalfa. Larval movement, mostly in the late instars, generally begins when the alfalfa has been cut and the larvae are forced to find food. If the armyworm population is high and the alfalfa shows signs of defoliation, the larvae will disperse before the field is harvested.

Description

Mature larvae are about 38 to 76 mm (1 1/2 to 3 inches) long. They are velvety black on the dorsum with two prominent and many fine bright yellow stripes on the sides. The underside is reddish. No other larvae found in alfalfa or grapes have these distinguishing features.

Injury

In outbreak years treatments are often needed to prevent serious defoliation in vineyards.

Seasonal Development

In alfalfa eggs are laid in masses that are covered with downlike scales, almost white in color. Upon hatching, the small blackish larvae remain and feed near the oviposition site. The first and second instars strip the lower epidermal layer from the alfalfa leaves. These "white cap" feeding spots are easily seen in alfalfa, most commonly on the irrigation borders. The later instars eat all of the alfalfa leaves.

Outbreak of this cyclical pest seems to be related to those years with above-average rainfall. Recent outbreaks occurred in 1969, 1971 and 1978, all years of heavy rainfall. Heavy rains prolong green vegetation in the foothills, permitting nearly all of the first generation's egg masses to develop to mature larvae. The emerging adults mainly fly to alfalfa on the valley floor and give an added boost to the normal summer generations.

In outbreak years roadways next to untreated mature or cut alfalfa (mainly in July and August) will be covered with millions of larvae. Roadways 40 to 50 feet wide are crossed with ease by the migrating larvae. Indeed, well-paved county roads often become slick as the tires of autos and trucks crush the dispersing larvae.

Natural Control

The western yellowstriped armyworm is attacked, mostly in the early instars, by a number of parasites and predators. Frequently a polyhedrosis virus will destroy nearly all larvae in an alfalfa field. Heavy virus kill has also been observed in grapes. The virus disease is more often seen during the fall on late instar larvae.

Monitoring Guidelines

Yellowstriped armyworm populations in adjacent alfalfa fields should be closely observed to determine potential influx in vineyards.

Management Guidelines

When massive numbers of larvae begin crawling from alfalfa fields, a V-shaped trench or small ditch should be cut alongside the vineyard. Contact insecticide should be dusted on the bottom of the cut in sufficient quantity so that the dust can be seen. Often migrations are so heavy and prolonged that the larvae crawling about the trench or ditch will move the dirt and cover up the insecticide. In other cases the migration will continue beyond the effectiveness of the insecticide dust, which is degraded by sunlight. In such cases repeat applications will be necessary

When mechanical barriers are not practical and if there will not be a residue problem in the vineyard or a drift problem on the alfalfa, ground or aircraft applications can be used when a migration is moving across the vineyard rows. Usually, treatment of eight to ten outside rows with a ground rig or one or two swaths with aircraft will be sufficient to eliminate larvae that have entered the vineyard. Aircraft applications are more practical when the migration is down

Grape bud beetle adult.

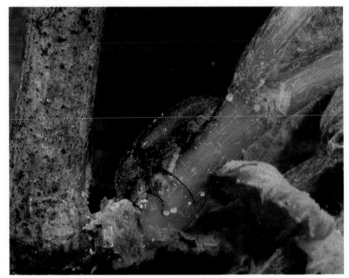

Branch and twig borer on grape shoot. See photo below for its effect 10 days later.

Damage to shoot by branch and twig borer seen in photo above. Arrow points out where adult burrowed into crotch.

Western grape rootworm adult, black phase.

Western grape rootworm adult, brown phase.

Photo by Amand N. Kasimatis

Berry damage caused by western grape rootworm feeding.

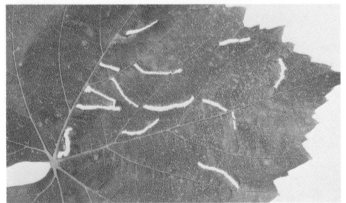

Feeding injury to leaves caused by adult western grape rootworms.

GRAPE BUD BEETLE

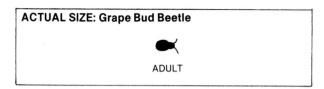

ACTUAL SIZE: Grape Bud Beetle

ADULT

The grape bud beetle, *Glyptoscelis squamulata* Crotch, is indigenous to western states. First noted as a grapevine pest in Nevada's Las Vegas Valley in 1922, it spread over California's entire Coachella Valley in the 1920s and became the major grape pest, sometimes destroying as much as 90 percent of a crop. Today, damaging populations are scarce in that area as the result of insecticide use for other pests, and annual treatments for grape bud beetle are no longer required.

In 1936 this species was first reported on grapes in the Kings River bottoms near Sanger, and it is now well established in parts of the San Joaquin and Sacramento valleys where it is considered a minor grape pest and is rarely a problem. Grape bud beetle also feeds on various weeds and is found on poplar and willow. Adults have also been observed feeding on the callus tissue of grafts on plums.

Description

Adults are about 6.4 mm (1/4 inch) long, hard-shelled and covered with light to dark gray pubescence.

Injury

Like cutworms, grape bud beetles feed on opening buds at night during spring. During the day they hide in the rubble on the ground or under loose bark on the trunk and arms of the vine. They feed on swollen or opening buds, usually starting at the tip of the bud and gouging out the heart, leaving the bud scales nearly intact. Close inspection is therefore necessary to discover the hollow, dead buds. After normal shoots are 1 or 2 inches long, the injured buds are conspicuous by their lack of growth, but by this time any significant damage has already been done. Usually the secondary and tertiary buds are not injured, and one or both of these may grow to produce shoots that are less fruitful than the primary.

Seasonal Development

In spring the female lays eggs in compact masses of 20 to 30 eggs each and conceals them in the deepest crack in the grape bark. In a few days the eggs hatch and the young larvae crawl or fall to the ground and immediately burrow into it. They then seek the roots of the vine and feed on them without, however, producing noticeable loss of vitality in the vine. The larvae may go two or three feet deep in quest of roots, remaining in the soil all summer and winter. In early spring the larvae construct smooth cells in the soil, in which they transform to the pupal stage from which adults later emerge.

Natural Control

No information is available on natural control of grape bud beetle.

Monitoring Guidelines

In some years grape bud beetles may be numerous in the vineyard and still remain unnoticed by the grower because, like cutworms, they conceal themselves during the day. From observations of damage alone, it is difficult to decide which pest is responsible. To satisfy curiosity, a grower can examine the vineyard with a flashlight after dark on a warm spring evening.

Management Guidelines

Only a few beetles feed on the buds at one time; the great majority remain hidden under the vine's bark. Consequently, if few beetles are seen, it should not be concluded that the infestation is so slight as to be of no economic importance.

Dusts or sprays have been used successfully for control. Applications should be made to the stakes as well as to the trunks and arms of the vine. The purpose is to cover with insecticide all paths the beetles may take from their hiding places to the swollen buds.

FLEA BEETLE

ACTUAL SIZE: Flea Beetle

ADULT

The flea beetle is so named because it can jump like a flea. Several different species are known to attack grapes in different parts of the country, but in California the steel blue, grapevine flea beetle, *Altica torquata* LeConte, is the only species of importance.

If nymphs are found moving onto vines, an insecticide should be applied at once to both vines and weed sources. Bugs migrate mainly in one direction and the wilted vines along the edge of the vineyard clearly show the line along which they are entering. It is possible to lay down a chemical barrier about 30 feet wide to prevent further migration.

EUROPEAN EARWIG

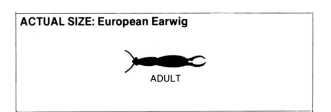

ACTUAL SIZE: European Earwig

ADULT

The European earwig, *Forficula auricularia* Linnaeus, is occasionally found in vineyards and may cause some damage to young foliage. The insect, an introduced species, was first discovered in Berkeley in 1923. Since then, it has become widespread in California and is most destructive in the cooler, coastal regions on many kinds of plants including fruits, vegetables and ornamentals.

Description

The European earwig is a dark, chestnut brown insect about 13 mm (1/2 inch) long with a pair of forceps on its body's posterior. It has a pair of short wings, but it usually moves about by running. The pearly-white, oval eggs are laid in masses in the soil. The young nymphs, with the exception of their small size and winglessness, are similar in appearance to adults.

Injury

Damage to grapevines caused by earwigs feeding principally on young foliage from April to July is generally confined to the basal leaves. These leaves may appear perforated or their margins may become ragged.

Seasonal Development

Earwigs overwinter as adults. Eggs are laid from December through February. The young nymphs and adults are nocturnal and come out at night to feed. During the day, they may be found among the leaves, in trash on the ground, or under rocks and in crevices. In the vineyard earwigs are often found under the loose bark of vines.

Monitoring Guidelines

Earwigs are usually not a problem in the vineyard. In early spring some damage may occur on young foliage, but vines generally are able to outgrow the injury. Earwigs may be observed when the loose bark near the damaged area of the vine is peeled off.

Management Guidelines

There are no recommended controls for earwigs on grapes.

VINEGAR FLY

ACTUAL SIZE: Vinegar Fly

ADULT

The names of vinegar or pomace fly are applied to various species of *Drosophila* but especially to *D. melanogaster* Meigen and *D. simulans* Sturtevant. These two species are difficult to separate and will be considered together. During harvest in vineyards and orchards in Fresno seven species of Drosophila can be found, of which more than 95 percent are generally *melanogaster* and *simulans.*

Description

The yellowish adult flies are well known because of their attraction to fermenting fruits of all kinds. They are about 2.5 mm (1/10 inch) long and are often seen hovering above garbage cans and cull fruit and vegetable dumps. The eggs have two appendages. The larvae are typically maggot-shaped and are about 6.9 mm (1/4 inch) long.

Injury

As grape berries ripen they may pull away from their stems, especially when clusters are tight. This exposes the fleshy part of the fruit and is an attractive spot for the laying of vinegar fly eggs. Hatching larvae then feed on the berries. Adult flies are also attracted to fermenting bunches, and as they fly about they carry bunch rot pathogens from infected bunches to previously uninfested clusters. Their greatest damage to the vineyard occurs with this secondary spread of bunch rot.

Seasonal Development

Few vinegar flies survive winter when fully exposed outdoors. They may survive and reproduce in masses of fermenting material where the inner temperature of the mass is favorable. They also survive indoors wherever fruit is stored. *D. melanogaster* has a life cycle of seven to eight days at 29.4°C (85°F), but the cycle may take 70 days during the winter. The females live an average of 28 days at 28.3°C (83°F) and lay a maximum of 26 eggs per day or 500 to 700 eggs in a 25- to 30-day life span. The minimum temperature for flight is 12.8°C (55°F) and the maximum is 37.8°C (100°F). Temperatures above 40.6°C (105°F) kill adults in a few minutes. Adults find their food by odor and thus fly upwind, although air movement greater than seven miles per hour restricts flight activity. Both adults and larvae prefer to feed on yeasts.

Vinegar fly populations build up during the growing season on culls and wastes of several fruit and vegetable crops grown in areas near vineyards. This buildup is slowed by hot weather, but if a sudden cool spell or a light rain occurs during harvest, huge populations develop quickly.

Natural Control

Natural controls are not applicable for the management of vinegar fly.

Monitoring Guidelines

No monitoring guidelines have been developed for vinegar fly.

Management Guidelines

Cultural management of fertilizer and irrigation programs and use of gibberellins (Thompson Seedless only) may reduce the number of tight bunches and incidence of bunch rot.

Chemical control of vinegar fly may reduce the amount of rotting bunches from 10 to 12 percent (by weight) below the amount of untreated vines. Treatment should be started when bunches start to show rot and become attractive to vinegar flies. In the Fresno area this should occur from mid- to late August.

Postharvest applications of pyrethrins effectively reduce flies on and around fruit in the field, in packinghouses and in storage and processing plants.

—WESTERN SUBTERRANEAN TERMITE—

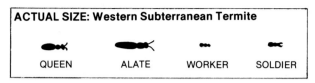

ACTUAL SIZE: Western Subterranean Termite

QUEEN   ALATE   WORKER   SOLDIER

The termite that attacks grapevines in California is the western subterranean termite, *Reticulitermes hesperus* Banks. Like all subterranean termites, *R. hesperus* feeds on wood but must maintain contact with the colony in the ground. On warm, sunny days following rains winged termites may be seen swarming from their subterranean galleries through small exit holes in the ground.

Termite damage to the dead heartwood of the living grapevines may go unnoticed until an arm or a whole vine is weakened by their feeding. Damage usually is not evident until the vines are at least 20 years old or older.

Description

Termites are social insects; their various castes include:

Prealates—immature wingless form; body is white; abdomen shrinks as wings develop.

Alates—winged reproductive that swarms; total length, including wings, is about 10 mm (2/5 inch); the body itself is black and about 4.8 mm (3/16 inch) long. As with all termites, the body is comparatively uniform in width. The wings are light gray. Unlike the winged ant, for which it is often mistaken, it has no slender waist.

Primary queen—large body (8 to 9 mm or 3/10 inch long) and wingless; wings break off after swarming takes place and the new colony is established.

Supplementary queens—a wingless form that never leaves the colony; lighter color than primary queen but somewhat darker than other castes.

Workers—white, eyeless, the body is about 3 mm (1/8 inch) long; feeds on and destroys wood and is the most abundant caste in the colony.

Soldiers—white body, about 5 mm (1/5 inch) long; head is armed with long jaws, is eyeless, quadrangular, about the size of the body, and is light colored. Soldier remains in the colony to protect other castes from insect enemies, especially ants.

Injury

Termites eat the heartwood (dead tissue) and avoid the living sapwood when they attack the vine. They will live for years in heartwood and only slightly penetrate the outer surrounding sapwood. Usually, the entire core of the heartwood is honeycombed, and its structural strength is so weakened that breakage occurs with the least unusual strain.

Old vines show more termite damage than do young ones. In some districts vineyards 40 years or older are infested with termites 100 percent because they have had much greater opportunity to become infested and their productive sapwood has become much thinner. Also surface injuries and heavy saw cuts expose more of the acceptable heartwood to possible termite invasion.

Seasonal Development

The winged termites swarm to reproduce and found new colonies. (All other caste members: prealates, workers, soldiers, queen and supplementary queens are wingless and are left in the old colony.) After flight a male and female break off their wings and found a new colony by excavating a shallow cell in the earth, usually under wood or vegetation. Eggs are laid in a "royal cell." Development of the primary colony is slow and the first generation of winged reproductives in this colony cannot be expected before three or four years at the very earliest. Supplementary queens eventually develop to build up the population of the primary colony.

Termite activity increases in warm environments, and for this reason termites are abundant in California.

As soon as the workers are able, they seek edible wood in the vicinity of their nest. Wounds exposing the grapevine heartwood, old beetle holes or unhealed pruning cuts are entry sites.

Natural Control

Natural control of termites is not applicable in vineyard pest management.

Monitoring Guidelines

No monitoring guidelines have been developed for termites.

Management Guidelines

Control is a matter of prevention. Care should be taken to avoid scarring the vines with cultivating tools. Saw cuts 12 inches or more above the ground are rarely a point of entry for termites, unless the heartwood is softened by wood rot fungi or reduced by branch and twig borers. In such cases the king and queen may establish their nest there as though it were soil.

A vine replacement program should be initiated in older vineyards. Severely weakened vines may be replaced by layers from adjacent healthy vines.

Stakes should be treated to protect them from wood-rotting fungi and termites.

———GRAPE ERINEUM MITE———

ACTUAL SIZE: Grape Erineum Mite

(CAN BE SEEN WITH A 14x HAND LENS
OR WITH A DISSECTING MICROSCOPE)

The grape erineum mite, *Colomerus vitis* (Pagenstecher), is an eriophyid mite. Widely distributed in California, it has been found on almost every cultivated variety of grape. Its only known host is the grapevine. Damage generally is minor in commercial vineyards. Although indistinguishable from each other, three different strains of the erineum mite have been recognized by the characteristic injury caused by each strain. These are the erineum strain, the bud mite strain and the leaf curl strain.

Description

The mite is best seen with a dissecting microscope. Adults are white, wormlike and about 0.2 mm (1/125 inch) long and less than 0.05 mm (1/500 inch) wide. Mites are gregarious, living in colonies within which may be found many of their castoff skins and their eggs, which are oval and whitish.

Injury

Erineum strain. Damage is first noticeable in early spring when infested young leaves show bright pinkish or reddish swellings or galls on the upper surfaces. The undersides, beneath the galls, are concave and densely lined with felty masses of curled plant hairs. These felty patches are called "erinea" (or "erineum" for a single patch). In a short time the swellings on the upper sides turn to a normal green, but later in the season erinea turn yellow and in August they turn brown. An infested leaf may reach a diameter of about two inches and die. Because the leaf remains attached to the cane, the mites can leave the erinea and move to new leaves; or a leaf with as many as 20 to 30 galls may expand to full size and appear normal, but matures and drops in the fall, somewhat earlier than a noninfested leaf.

Bud mite strain. These mites live within the grape buds and do not live in or produce erineum on leaves. They feed at the base of the outer bud scales and cause blisterlike growth on the inner surfaces. Generally confined to the outer bud scales, they may penetrate deep into the bud and feed on embryonic tissues of the shoot primordia. Common symptoms: short basal internodes, scarification of the bark of new shoots, flattened shoots, dead terminal buds on new canes, witches' broom growth of new shoots, zigzagged shoots and dead overwintering buds. Leaves are usually stunted and wrinkled; their veins are prominant and drawn together. Premature dropping of flower clusters may also result from injury to the immature leaf and flower buds. Closely similar symptoms are also caused by temporary boron deficiency in early spring (see UC publication 4087, *Grapevine Nutrition and Fertilization in the San Joaquin Valley*).

Leaf curl strain. Symptoms appear in summer and show downward curling or rolling of leaves. Severity of the rolling ranges from slight to severe curling so that the leaf tends to roll into a crude ball. Overall growth of the shoot may be stunted and scarring may sometimes be noted on the shoot. Lateral shoot growth may also be present.

Seasonal Development

Erineum mite. Adults overwinter under the outer layers of scales on dormant buds. They do not lay eggs during winter. As soon as buds open in spring, adults migrate to the unfolding leaves. Mites behave gregariously and are associated in small groups during formation of their first erineum. They live within the erinea and feed only on the lower leaf epidermis between the plant hairs. Abnormal growth and production of the mass of plant hairs is somehow stimulated by the mites feeding on the leaf surface. Some may move to other areas of the leaf or to other leaves and start new colonies. From mid-August to leaf drop the adults migrate back at night to the axils of the leaves and crawl under the bud scales.

Bud mites. These live within the buds and not on the leaves. Mites overwinter, mostly as adult females, and from December to late March the population declines primarily because of the gradual death of the overwintering adults. During spring and summer mites increase, penetrate deeper into buds and infest other buds further out on the shoot. The mites crawl from bud to bud or are carried during the shoot's growth. Infestation of buds on a cane is not uniform, but generally it is more prevalent in the basal buds. Females lay one egg a day, incubation is approximately six days, and the adult stage occurs about 14 days after egg hatch.

Leaf curl strain. The life cycle is not known, but it is assumed to be similar to that of the erineum strain.

Natural Control

The predaceous mite, *Metaseiulus occidentalis* (Nesbitt), is effective in reducing the erineum mite population. This predaceous mite overwinters under bark or under bud scales. Its population fluctuates considerably and is influenced by the abundance of prey, including the erineum mite present in the fall, and by the extent of winter mortality. In spring expanding foliage exposes the erineum mites, which are particularly susceptible to the predaceous mites at this time. Because of its larger size, the adult predaceous mite may be restricted in its movement by the mass of plant hairs in the erineum, or it may not be able to penetrate deep inside the bud scales in compact buds. Immature stages, because of their smaller size, can gain access and feed on the erineum mites.

Monitoring Guidelines

Economic loss from either the erineum or leaf curl strains is light. In the case of bud mites the frequency of buds damaged severely enough in a vineyard to cause commercial loss has not been observed.

Erineum mite. Look for the characteristic symptoms on young leaves early in the growing season: red galls or swellings on the upper leaf surface and portions of the leaf's underside lined with erinea. Microscopic examination of the erineum will reveal small, white, wormlike mites at the base of the plant hairs.

Bud mites. When symptoms associated with this mite appear, the basal buds should be cut off the cane and the outer bud scales opened and examined under the microscope. Infested bud scales will show blisters and scars on the inner surface with the presence of mites, in contrast to the smooth inner surface of normal bud scales.

ACTUAL SIZE: Sawtoothed Grain Beetle				
	first	mature		
EGG	LARVAL STAGES		PUPA	ADULT

The sawtoothed grain beetle, *Oryzaephilus suri-namensis* (Linnaeus), a member of the family Cucujidae, is a cosmopolitan pest that feeds on practically any stored dried food. Fairly dry foods are preferred. Both larvae and adults attack the commodities. The sawtoothed grain beetle infests all cereals (rice, wheat, maize, barley and pastas such as macaroni), bread, flour, nuts, copra, starch, drugs, tobacco and dried fruit. Raisins are one of its favorite foods. In raisins stored for a year or more this insect can become abundant. It crawls rapidly, even on vertical surfaces, but it has not been observed to fly. Newly hatched larvae can enter extremely narrow crevices in search of food.

Description

The adult beetle is about 3 mm (1/8 inch) long, narrow and flat. It is brown and has six toothlike projections along each side of the body in front of the wings. A look-alike beetle found in the San Joaquin Valley is the merchant grain beetle, *O. mercator* (Fauvel). For an excellent comparison, see figure 26 of *Stored-Grain Insects*, Agricultural Handbook, USDA No. 500, February 1978.

The eggs of the sawtoothed grain beetle are white, elongate-oval and less than 1 mm (1/25 inch) long; they are not visible to the unaided eye. The larvae are yellowish-white and when fully grown are about 3 mm (1/8 inch) long. Pupae are white or yellowish-white and are found in or near larval food. The larva usually makes a cocoon for protection before pupation. The cocoon consists of fine particles of food cemented together by the larva.

Injury

This beetle attacks all parts of the raisin, feeding as much in the deep folds as on the ridges. No webbing is deposited as is the case with Indian meal moth or raisin moth. The excreta are yellowish pellets, more elongate than those of the raisin and Indian meal moth larvae, and similar in size and shape to the beetle's eggs.

Seasonal Development

Time from egg to adult is 27 days in summer. Females begin to lay eggs in about five days after pupal emergence. From six to ten eggs are laid each day singly or in small clusters until they total from 45 to 285 eggs. They are laid in crevices formed by tight folds in the raisin skin, and the hatch takes from three to five days in midsummer and from eight to seventeen days in spring and fall. In summer, the larvae develop in about two weeks, but in spring they feed and grow for four to seven weeks. During larval development they molt two to four times. Under certain conditions the larvae are cannibalistic, and some authors believe Indian meal moth populations may be suppressed by large populations of sawtoothed grain beetle. Pupal period is six to nine days in summer.

Sawtoothed grain beetles are long-lived. Some have been recorded as having lived for more than three years. Optimum survival temperatures are between 30° and 35°C (86° and 95°F). There are normally five to six generations each year, but indoors in warm buildings breeding and development occur throughout the year. Thus, infestation may spread during storage. In California's Coachella Valley adults and larvae remain active outdoors throughout the year in dropped fruit in date groves.

Natural Control

A parasitic wasp, *Cephalonomia tarsalis* (Ashmead) of the family Bethylidae, has been found to reduce sawtoothed grain beetle populations by attacking the larva. This wasp will not economically control an infestation.

Monitoring Guidelines

Screening an occasional sweatbox or composite sample from many sweatboxes usually will confirm the presence of any insects and the need for fumigation.

Sawtoothed grain beetle larvae.

Management Guidelines

Control of this pest (and other pests of raisins) is accomplished primarily in the packinghouse. Farmers who cover and make sweatbox stacks or bins on the ranch should have fumigation programs. The first fumigation should be made when the stack is completed, and the stack should be fumigated again periodically when storage exceeds 60 to 90 days.

As raisins are removed from storage they undergo several processing steps before being packaged for sale. Most processors subject raisins to: shaker screens, sorter-blowers, cleaners, washers, rifflers and stemmers. By the time the raisins are packaged for consumer sales, immature, damaged, lightweight or otherwise undesirable fruit has been removed.

The atmospheric control mentioned in the section of this publication on the driedfruit beetle would be effective against this insect.

Sawtoothed grain beetle adult. Thorax of an adult is seen in closeup.

NEMATODES

Photo by Michael V. McKenry

Section VI—NEMATODES

Contents

On the preceding page:

The aerial photo shows weak spots in vineyards caused by nematodes. See page 243 for details. The photo below, taken from the top of a pickup truck, shows stunted vines which possibly have been affected by nematodes.

Nematodes are microscopic, multicellular, nonsegmented roundworms commonly present in soil and are adapted by the structure of their mouthparts to derive nutrients from either: (1) soil microorganisms such as bacteria, fungi or other nematodes or (2) from plant roots.

Communities seldom exist as a single species and both parasites and nonparasites of grape occur in most vineyard soils. A typical vineyard soil contains nematodes feeding on grape roots, on other organisms associated with grape roots or other biological components of soil and on other vineyard plants or weeds.

Plant parasitic nematodes reduce root efficiency. Vine damage is eventually manifested as reduced vigor and yield with slight yellowing of leaves. Vine death seldom occurs unless there are other stresses on the plant. These symptoms are common for other root-restricting agents.

Roots of nematode-infected vines are unable to meet aboveground demands for nutrients and water, especially during peak demand periods, and are the first to show a nitrogen or water deficiency. On nematode-tolerant grape varieties, vine stress can be minimized by regulating cropping loads with extensive pruning. Nematode damage and symptomology are nonspecific compared with most aboveground maladies.

Because aboveground vine symptomology is lacking and nematodes are microscopic in size, laboratory analyses are necessary to determine population levels and to identify species. Good analyses depend on use of proper techniques by the grower, vineyard manager or consulting laboratory to take soil and plant tissue samples in the field. Sampling methods are outlined at the end of this section.

During recent decades studies of soil fumigation, nematode taxonomy and soilborne viruses have enhanced understanding of the belowground arena in vineyards, especially the nematode components. As viticulturists promote vine productivity and longevity, healthy root systems become more significant. Similarly, it is necessary to consider aboveground conditions when dealing with soil problems.

It must be recognized that methods for controlling nematode populations can upset the total microflora and fauna of soil. For example, use of a sod cover crop may minimize buildup of root knot nematode, but new species of root lesion nematode will likely flourish; the nematicide DBCP was commonly reported to reduce all nematode populations, but stubby root and pin nematodes frequently increased following its use; the preplant fumigants methyl bromide* and 1,3-D*, when properly applied, profoundly affect nematodes, but populations frequently recolonize quickly following incomplete fumigations.

It is of primary importance, therefore, to get a good analysis of the species present and the population levels. Then in conjunction with other vineyard factors, such as soil and general plant conditions, one must deal with individual and combined effects on plant growth and crop production.

Prohibition of the use of DBCP has promoted interest and research in affirmative control agents and methods but it would be premature to discuss them as of this writing. We will, however, allude to several vineyard management decisions which influence the degree of economic loss by nematodes.

Following are descriptions of nematodes damaging to grapes and most commonly found in California vineyards. They are listed in the order of their frequency. Included are data on population dynamics, geographic distribution, symptomology and injury. Information is not provided on nematodes which are not parasitic on grapes, even though they may feed on weeds or other vineyard plants.

ROOT KNOT NEMATODES

Root knot nematodes include *Meloidogyne incognita, M. javanica, M. arenaria, M. hapla* and other *Meloidogyne* spp., some of which remain unnamed. From a management standpoint, it is currently important only to determine that root knot nematodes are involved, not the particular species. Second stage juveniles hatch from eggs and move through moist soil to contact roots. Vertical distances of one foot in sandy loam soil can be traveled in three days by those most active.

Description

The active juveniles are attracted to roots and usually penetrate and enter just behind the root tip. Once inside the root they establish themselves in the conducting tissues, begin feeding and after two weeks of

---

*Restricted material; permit required from County Agricultural Commissioner for possession and use.

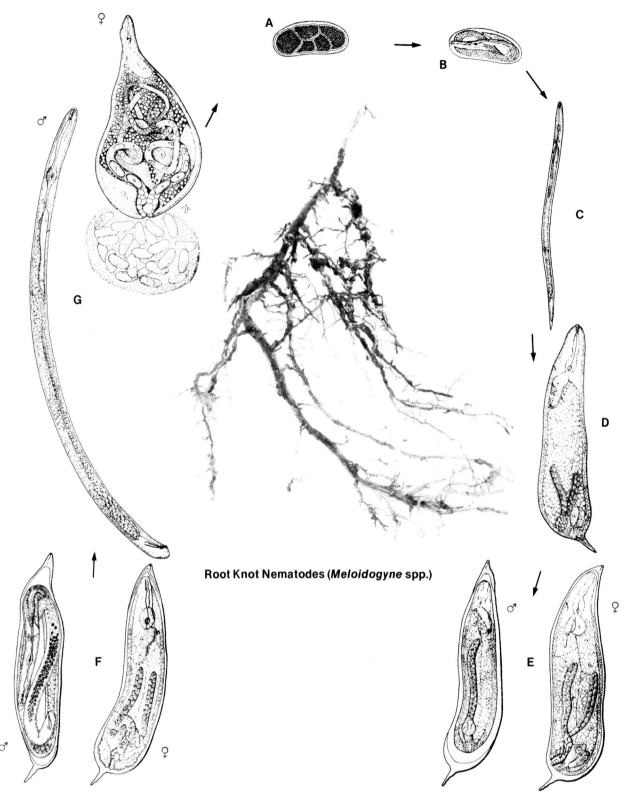

Root Knot Nematodes (*Meloidogyne* spp.)

The life cycle of a sedentary endoparasite, root knot nematode, is depicted: (A) Egg stage. (B) First stage juvenile within an egg. A molt occurs within the egg. (C) A second stage juvenile emerges from the egg. The second stage juvenile must penetrate the root to feed. (D) Once inside the root, the second stage enlarges. (E) Third stage male and female. (F) Fourth stage male and female. (G) Adult stages. The vermiform male does not feed. The saccate female remains sedentary, producing eggs in a gelatinous matrix which usually develops external to the root surface. The life cycle is completed.

Other nematode species have slightly different life cycles. Root lesion nematode, a migratory endoparasite, remains vermiform in the adult stages and either the second, third, fourth or adult stages may penetrate or vacate a root. Stubby root nematode, ring nematode, needle nematode, pin nematode and dagger nematodes have life cycles similar to the root lesion nematode, except that they feed only at the root surface and do not enter roots.

warm, summer temperatures the females mature into egg laying adults. Their development stimulates a cellular change in the plant in the immediate vicinity of the feeding site. This change results in formation of the familiar "knot" or "gall" seen on the root surface. Internally this results in disruption of conducting tissues. A single gall may be inhabited by one or numerous adult females. The number of females living in a single gall apparently influences its size. The adult female is a sedentary, pearl-colored stage, which, if dissected from the gall, is barely visible to the unaided eye.

The life span of root knot nematode in grape presumably is from one to several months, with the greatest longevity but least activity occurring during winter. It is apparent that a single gall can be maintained by successive females for many years.

Root knot nematode males, while sometimes present in low numbers, do not feed and are not of direct concern to the grape grower.

The egg population reaches highest numbers in September when soils of many vineyards are driest. Presumably, dry soils reduce egg hatch but not egg production. As many as 1500 eggs may be produced by a single adult female in a Thompson Seedless root. Second stage juveniles in soil are generally two to five times more numerous during fall and winter than in spring and summer.

Surveys indicate the preponderance of the population is located 6 to 36 inches deep beneath the vine row, depending on soil conditions and tillage practices, but wheel traffic and its effect on soil compaction and root distribution minimize this nematode's development in the drive row.

Root knot nematode is best adapted to coarse-textured soils including sand, loamy sand and sandy loam. It exhibits a wide host range including the roots of many broadleaf weeds and cover crops present in vineyards.

Most prevalent in cooler regions north of San Francisco, *M. hapla* has been found in southern California vineyards, perhaps being more active there during winter months.

Meloidogyne incognita, M. javanica and *M. arenaria* are most prevalent south of the Livermore Valley and Modesto areas. Root knot nematodes are less prevalent in the north and central coast valleys.

Injury

Root knot nematodes interfere with plant growth and nutrient uptake. They create "sinks" in the root system that channel aboveground photosynthates to them. They further disrupt the orderly uptake of water and nutrients by their physical presence within root tissues. Damage caused by them is increased if plants are stressed.

Small galls or knots, usually present on infected grape roots, are typically 1/8 inch in diameter, but can be larger where there has been a multiple attack. However, they are seldom larger than 1/2 inch in diameter, even on older roots. Root knot nematode confirmation requires examination of the gall for the presence of adult females.

DAGGER NEMATODE

Xiphinema americanum, the most common species of dagger nematode, is an external parasite generally known for its larger size and lengthy root-penetrating spear that allows it to feed deep in root tissues. It prefers young roots of woody plants, but will reproduce and feed on diverse plants including sudangrass, strawberry and alfalfa.

Description

Juvenile stages are most evident during early spring months. Population levels are approximately twice as high in winter as in summer months. In vineyards with moderate wheel traffic, 85 percent of the vineyard population resides within the surface 18-inch soil zone directly beneath the vine row. It is thought that the life expectancy for this pest may exceed two years. Numbers of this species are reduced in zones where there is frequent tillage.

Except for the Coachella Valley, this nematode may be found in vineyards throughout California. There are also vineyards where it is conspicuously absent. Preference for a particular soil texture is not apparent.

Injury

Weakened vines in a high producing vineyard have been shown to yield correspondingly less as the population of this nematode increases. However, the poorest vines typically are also associated with other soil pests including root knot nematode.

This nematode and several other root parasites feed near the root tip and may cause an enlargement there. Since this symptom is not always present, it is not a reliable diagnostic tool for dagger nematode unless soil samples are also taken.

Xiphinema americanum is the specific vector of yellow vein virus.

ROOT LESION NEMATODE

Root lesion nematode (*Pratylenchus spp.*) is a common name for a group that includes more than 35 separate species. This endoparasite migrates through root tissues, resulting in root dysfunction.

Root lesion populations frequently achieve high levels in grassy vineyards; these types, however, are not considered to parasitize grape roots. The species are differentiated by their host range and morphological and anatomical proportions.

Description

Pratylenchus vulnus is the most important of this genus on grapevines. Soil populations are characteristically low in number. These nematodes migrate in and out of roots and females lay eggs within roots or in the soil.

Pratylenchus vulnus can be found throughout California but it is not uniformly distributed.

Injury

In young vines, top growth is noticeably restricted and vine recovery seldom occurs. Among older vines damage caused by this nematode is believed to be high, but data to support this are incomplete.

The root system of young vines planted in soil infested with *P. vulnus* and previously supporting a perennial crop may be severely restricted. The best symptom is the lack of roots on affected vines, coupled with a nematode sample to confirm its presence. Dark colored lesions on the root surface are sometimes found, but this symptom is unreliable as a diagnostic tool.

CITRUS NEMATODE

The nematode, *Tylenchulus semipenetrans,* has a narrow host range. In California agriculture it is limited to citrus, persimmon, olive and grape. While only one species is involved, there are at least five different races. This nematode attacks behind the root tip, becomes sedentary and, except for its embedded neck region, stays outside the root.

Description

Populations of citrus nematode develop to high numbers in vineyards. This nematode favors loam-type soils including loam itself, sandy loams and clay loams. On shallow soils citrus nematode achieves high levels in the drive row as well as in the berm area.

Citrus nematode is most prevalent along the San Joaquin Valley's eastern side, including areas east of Porterville, Delano, Tulare and Fresno. It is present in vineyards several miles from citrus groves and in vineyards with no history of citrus.

Injury

Citrus nematode does reduce vine vigor and yield. The best evidence of field damage has come where growers have had improved plant response after using DBCP in vineyards where it was the predominant species.

Root samples taken from soils with high populations frequently have soil particles clinging to the root surface, even after a vigorous shaking. This is because of soil embedded in the exposed egg matrix. This "dirty root" symptom is not a reliable diagnostic tool, but where nematode analyses have verified its presence, it can help determine the extent of field infestations.

PIN NEMATODE

The ectoparasite, *Paratylenchus hamatus,* the smallest of the plant parasitic nematodes, has a long stylet allowing feeding deeper than the epidermis. It prefers the roots of woody plants and exhibits great preference for Nemaguard peach with a relatively moderate preference for grape roots.

Description

Pin nematode can build to relatively high population levels. Fourth stage juveniles are a resistant stage capable of withstanding treatment materials (including applications of DBCP). This species is as prevalent at the three-foot depth as at the one-foot depth.

Pin nematode is especially common in southern portions of the San Joaquin Valley south of Merced.

Injury

Research has not shown any reduction in yield or quality of grapes because of this nematode's presence. In fact, one study indicated that *P. hamatus* populations tended to be higher among highest yielding vines. No symptoms are visible.

Section VII—VERTEBRATES

Contents

Rodents and rabbits, as well as some other mammals, are found in many California vineyards, and in certain situations they can cause considerable damage, resulting in lower yields and, not uncommonly, death of entire vines. They also interfere with cultural operations, such as irrigation, and in general vineyards should be routinely monitored to detect potentially damaging populations.

Clean cultural practices—elimination or reduction of heavy weed cover in or near the vineyard—contribute to fewer rodent problems. Weedy ditches, fence lines, adjacent fields, pastures and brush or trash piles are excellent reservoirs for many field rodents and can contribute significantly to problems.

Several factors must be considered for a successful rodent control program: (1) Identify the species causing the problem. (2) Alter the habitat, if possible, to make the area less favorable to the pest. (3) If population reduction is necessary, use the control method appropriate for the location, time of year and other environmental conditions. (4) Establish a monitoring system to detect reinfestation so that you can decide when control is necessary.

Rodent control equipment (baits, fumigants and traps) is generally available at farm supply and hardware stores, nurseries and garden shops. In addition, the County Agricultural Commissioner can sometimes provide rodent control materials. If needed materials cannot be located or more information is needed, consult the local county farm advisor or commissioner.

Pocket Gopher

Pocket gophers, frequently encountered in vineyards, are active throughout the year and, if uncontrolled, they will increase to high numbers. They can cause considerable vine damage and loss during the vineyard's lifetime as well as damage to irrigation and other cultural operations

Description

Stout-bodied and short-legged, these rodents have two pairs of prominent incisor teeth. External fur-lined cheek pouches open outside the lips on each side of the mouth and are used for carrying food and nesting materials. The head and body measure about 175 mm (6 to 8 inches) and the scantily haired tail is less than half the body length. The color ranges from yellow-brown to gray-brown. Weight is about 100 grams.

Biology

Five species of pocket gophers (*Thomomys* spp.) are found throughout California, except in rocky outcrops, high mountain meadows and some desert areas. They are most common where ample moisture and good soil encourages abundant plant growth. Named for their cheek pouches, they feed primarily on succulent underground parts of plants, but they do pull entire plants underground. At certain times they graze on plants aboveground near their burrow openings.

These gophers live underground, except when the young leave the nest after weaning to search for new places to live. They are antisocial and solitary, except during breeding and when the young are being raised. Burrow systems include main tunnels, side tunnels to push out dirt and characteristic soil mounds. Main burrows are normally 4 to 12 inches under the surface, but they can be deeper. Burrow openings are plugged with soil so that the system is completely enclosed, stabilizing the burrow's temperature and humidity at close to optimal conditions.

In uncultivated and unirrigated areas they normally breed after rains begin and green forage is plentiful; typically this results in one litter per year. On cultivated lands green forage all year long allows them to breed up to three litters per year. Litters average five young with a range of one to thirteen. After weaning, the young are expelled from the burrow to find new territories. Pocket gophers are active all year and may reach densities of 50 per acre. Legumes such as alfalfa and clover are preferred.

Legal Status

Pocket gophers are classified as nongame mammals by the California Fish and Game Code, and when they are found to be injuring growing crops or other property, they may be taken at any time or in any manner by the owner or tenant of the premises. The code specifies certain conditions to be met if leghold, steel-jawed traps are used.

Injury

There are a few plants gophers will not eat, but they prefer fleshy-stemmed and bulbous-rooted plants, especially legumes, more than fibrous species such as grasses. Damage to grapevines results when gophers cut roots or gnaw bark from the roots or trunk. Vines are commonly completely girdled a few inches below the soil line in a relatively short time with substantial economic loss resulting in newly planted vineyards in particular. Their burrows divert irrigation water, sometimes causing extensive soil erosion. They are known to gnaw and damage plastic irrigation pipe.

Damage by pocket gophers may be distinguished from that of rabbits and meadow mice because it normally occurs several inches below the soil surface, where they usually remove bark tissue down to the cambium layer. In contrast, meadow mice generally start feeding from the soil level upward for several inches, and often leave bits of bark attached to the wood. Gnawing marks of these animals are distinct and may aid in identification of the damaging species. Because damage is frequently not visible, it often goes undetected until a vine exhibits stress and is beyond help.

Natural Control

A variety of predators utilize pocket gophers for food, and they should be encouraged when they do not present problems (e.g., to livestock producers). It is generally accepted, however, that they do not keep populations in vineyards low enough to prevent economic injury.

Monitoring Guidelines

Population buildup in a vineyard is generally gradual and can be detected by an increased number of mounds.

Economic injury and treatment levels have not been established in vineyards, but even one gopher can destroy several vines, especially when other preferred vegetation is not available. Determination of the need for control depends on grower experience. As significant vineyard damage can result from very low infestations, control is recommended as soon as fresh mounds are detected.

Management Guidelines

Pocket gophers can be controlled effectively and even eliminated from vineyards by continuous, persistent effort. Preferred are poison baits, applied either mechanically or by hand, and trapping. Such other methods as repellents or fumigation with poison gas cartridges or automotive exhaust are not generally effective or recommended.

Toxic baits. Toxic baits placed in the burrow system are the most widely used. In hand baiting 0.2 to 0.5 percent strychnine* grain baits are commonly used. For machine baiting 2.6 percent or lower concentrations of strychnine on grain are generally used. Follow label instructions carefully.

Mechanical baiting. The mechanical bait applicator, a once-over operation, offers excellent control over large areas. This tractor-drawn device constructs an artificial burrow beneath the soil surface and deposits poison grain within the burrow at preset intervals and in preset quantities. It is ordinarily operated down any row where gophers are present. The artificial burrow will intercept most natural burrows, allowing gophers to find and consume the bait.

Several commercial manufacturers now build mechanical gopher-bait applicators, all of which operate on the same basic principle. All consist of the same four basic components on a supporting frame: A depth-adjustable, burrow-forming shank; a rolling coulter to cut surface trash and shallow roots ahead of the shank; a bait-metering device and a presswheel to drive the metering unit and to close the knifelike slit made by the shank's upper portions. The bait is dropped into the artificial burrow through a tube built or cast into the rear portion of the shank. (Soil type affects wearability of the shank; sandy and sand-clay soils cause the most abrasion.) Machine instructions generally state the rate most suited to the locality.

One man with the mechanical bait applicator and a wheel-type tractor (at least 25 horsepower) can treat from five to eight acres per hour. Tractor speeds of 2 1/2 to 3 1/2 mph (220 to 308 feet per minute) are commonly used; slightly faster speeds are possible with some sturdier models. Formation of a smooth burrow that does not collapse is important to good control and should not be sacrificed for speed—a point requiring special attention in sandy or light soil.

While learning the machine's operation it is desirable to open a small section of the artificial burrow with a shovel to inspect its depth and condition. A few trial runs may be necessary to properly adjust the machine.

Soil conditions: The machine can be used where soil condition and physical aspects of the land permit

---

*Restricted material; permit required from County Agricultural Commissioner for possession or use.

This mechanical bait applicator drawn by a tractor creates artificial burrows and deposits poison grain to attract one vineyard nuisance, the pocket gopher.

Photo by W. P. Gorenzel

formation of a good artificial burrow eight to ten inches below the soil surface. For satisfactory results the soil should be reasonably firm below the top three or four inches and be moist enough so that smooth, clean burrows are formed. Well-formed burrows are important in achieving control. Adequate soil moisture is necessary; in general, soil moisture should be suitable for good plowing or cultivating. A handful of soil, when squeezed, should retain its shape, yet not be sticky wet. Soils too wet tend to ball up the presswheel or cause loss of traction. In soil too dry the burrow will cave in. On irrigated land the machine can be used at any time when soil conditions are proper; in extremely sandy soils this is only a day or two after irrigation. After treatment, wait at least 10 days before disturbing the soil or a treated field with any other implement.

Artificial burrows need not be continuous. Short breaks caused by raising the shank to clear buried obstructions, such as irrigation lines, or to free the shank of trash will not adversely affect control. Only areas with gophers present will need treatment. Adding a burrow or two around an entire vineyard may help reduce gopher reinvasions from adjacent areas.

Burrow depth: The depth of an artificial burrow should average eight to ten inches, depending on the depth of the natural burrows of active gophers. This depth can be measured and the machine adjusted accordingly. For formation of a smooth open artificial burrow, however, the depth should never be

less than six inches. As depth is less important than a smooth and clean burrow, it is better to set the burrowing shank too deep than too shallow. Numerous obstructions, such as rocks and vine roots in the plow zone, may occasionally limit use of this machine.

Application rate: The bait application rate within the artificial burrow, usually expressed in pounds per 1,000 feet of burrow, is determined by: (1) the size (number of kernels) of each bait drop and (2) the intervals (distance) between bait drops. Sufficient quantity of toxic bait assures easy discovery. The distance between drops is generally between 18 and 48 inches, depending on the desired application. Instructions for presetting the machine to apply the desired rate are usually provided. Where there is no information on the preset rate of the bait applications, the machine may be checked for the desired calibration by placing a weighed amount of clean (unpoisoned) grain in the hopper and making a sample burrow for 1,000 feet. Weigh the grain remaining in the hopper to determine the amount applied; make adjustments as necessary.

At 20-foot burrow spacings 0.75 to 1 pound per acre of 2.6 percent strychnine* wheat bait is adequate for good control (0.34 to 0.46 pound per 1,000 feet). If the distance between burrows is closer, then the rate per acre will be proportionately greater and vice versa. Where locally recommended baits containing less than 2.6 percent strychnine are used, rates may be increased but generally these will not exceed four pounds per acre at 20-foot spacings (1.84 pounds per

*Restricted material; permit required from County Agricultural Commissioner for possession or use.

1,000 feet of burrow). Consult a bait distributor for local application recommendations.

The rates of application differ for different numbers of gophers suspected of being present. In instances of high populations, where inadequate control is achieved on the first treatment, a second application may be made, although it is advisable to wait at least two weeks before re-treating.

Hand baiting. A pointed 1/4-inch steel rod is used to find the burrow by probing near fresh mounds. The probe opening is then enlarged with a larger rod or broomstick and a small amount of grain-type bait is placed in the burrow. Then the hole is covered to exclude light. Metal probes with automatic bait dispensers save time and effort.

Trapping. To set traps which are effective in small areas or where only a few pocket gophers are present, find the main runway—one going in both directions. Probe with a pipe, rod or sharp stick a short distance in front of the low side of a fresh mound or follow a lateral runway from the mound. Dig a hole, clean out the burrow and insert two traps in the runway, one facing each direction. The traps should be wired to a stake so that they will not be lost. Cover the hole with a clump of sod or a board to keep soil from falling inside, and then sift loose soil around the sod or board to exclude all light.

Herbicides. Use of herbicides can reduce gopher populations by decreasing food supply. Clean cultivation also helps early detection of mounds. Controlling gophers in areas adjacent to the vineyards is necessary because young ones will disperse from there into the vineyard.

Irrigation. When vineyards are flooded, gophers may be drowned or forced to the surface where they are susceptible to human and natural predation. Normal

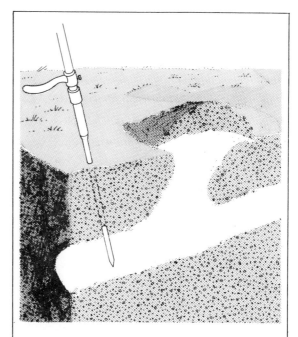

Probing to locate gopher's main runway.

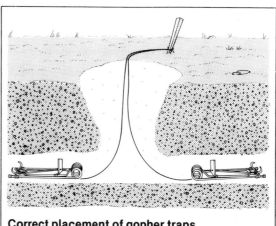

Correct placement of gopher traps.

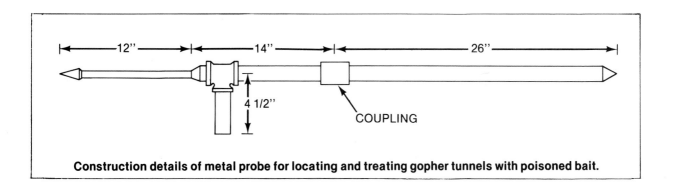

Construction details of metal probe for locating and treating gopher tunnels with poisoned bait.

furrow, sprinkler or drip irrigation does not detrimentally affect them because the water percolates around, not into, their tunnels.

Fencing. Although seldom used, a cylinder of wire netting (one-inch mesh or smaller) about twelve inches in diameter and eighteen inches tall can be sunk in the hole around a vine at planting to protect it against gophers. The top of the wire should extend at least six inches above the ground surface.

Repellents. No known pocket gopher repellents are effective for protecting grapes or other plants.

Meadow Vole

Meadow voles, *Microtus* spp., also called field or meadow mice, feed more readily on young vines which sustain greater damage than do older ones. During seasons favorable to meadow voles, damage can be severe enough to kill many vines, especially in young plantings that provide a favorable habitat. Populations develop in vineyards or on their borders and fencelines where grass, brush and trash may accumulate, and damage is most severe where grass or cover crops build up around vines to provide them with food and shelter.

Description

Blunt-nosed and short-eared, these rodents have small eyes, short legs and rather short tails. They have relatively coarse fur, usually dark or grayish-brown in color. Fully grown, they are larger than a house mouse but smaller than a rat; total length, including tail, is 170 mm (about 7 inches). The tail is 50 mm long.

Biology

Extremely prolific, the female may produce from five to ten litters a year, and within 15 hours after her young are born she may breed again. A few will breed anytime, but they principally breed during rapid forage growth. Populations fluctuate dramatically: their numbers usually peak about every four years, then fall abruptly to relative scarcity. At peak abundance there may be 1,000 to 3,000 voles per acre, each eating its body weight of 60 to 70 grams of green vegetation daily.

Legal Status

Like pocket gophers, meadow voles are classified as nongame mammals by the California Fish and Game Code.

Injury

These rodents feed on vegetation including stems, roots, bark and seeds of many plants. They damage or kill grapevines by eating through the bark to the cambium layer. Characteristic damage is complete or partial girdling of the trunk from just below the soil line up as far as they can reach on the trunk.

Natural Control

Clean cultivation and weed control on fence rows, roadsides and ditch banks are important preventative measures. Such areas are reservoirs where wintering vole populations may live, reproduce and migrate to adjacent vineyards.

Predators, such as coyotes, foxes, badgers, weasels, owls and gulls, may feed upon meadow voles, but they rarely, if ever, are a major factor in controlling a rapidly growing population.

Monitoring Guidelines

Most damage to vines occurs in winter or early spring, so monthly inspection then of vineyards and surrounding fields is essential for spotting vole activity and population increases, especially in heavily vegetated areas where new vole runways and burrows may indicate feeding.

Usually found in areas marked by numerous one- to two-inch-wide surface runways through matted grass, voles are active all year irrespective of weather and are generally most active during the day. Deposits of brownish feces and short pieces of grass stems along their runways indicate recent activity. Burrows are short and relatively shallow, have numerous round openings and frequently contain nesting and storage chambers.

The number of runways, the amount of freshly cut vegetation and the amount of fresh droppings found in runways indicate the number present. Any sign of activity calls for measures to prevent vine damage.

Management Guidelines

Vegetative cover provides both food and protection from predators. Clean culture with annual French plowing and middle discing or herbicide use can prevent cover buildup. Also, adjacent properties which may be harboring these pests will require cultural controls. Once voles are detected, toxic baits are generally the only way to eliminate them.

Chemical repellents have been tested, but none has proved effective in preventing vole damage to vineyards.

Toxic baits. Poison baits, to be most effective, must reach the voles' runways where most feeding occurs. Bait distribution may be done by spot baiting (placing bait by hand in runways and burrow openings) or by broadcasting (scattering bait over the entire infested area). Broadcasting may be done by hand, machine or, in limited situations, by aircraft. Broadcast application rates vary, depending upon estimated density of the vole population and the type of toxicant. Consult product label for correct application rates. The County Agricultural Commissioner is usually experienced in meadow vole control and should be consulted regarding the type of bait and proper application rate.

Single-dose baits: Zinc phosphide,* an acute rodenticide used for vole control in vineyards, is generally effective when used according to label directions.

Multiple-dose baits: Some County Agricultural Commissioners have an anticoagulant bait registered for meadow vole control. This material should be scattered lightly in tablespoon amounts (1/4 to 1/2 ounce) near active burrows or in runways. Repeat treatment every other day for three treatments. Commercial anticoagulant baits are also available. Follow label instructions when using these, as well as rodenticides.

Trunk guards. Young vines can be protected with cylindrical wire guards (which also protect against rabbits). They are made of 1/4- to 1/2-inch mesh hardware cloth, 24 inches wide and with a diameter sufficient for several years of uncrowded vine growth. To prevent voles from burrowing under them guards must extend about six inches below the soil surface. Meadow voles rarely climb these guards. Trunk guards made of plastic, cardboard and other fibrous materials, although sometimes less expensive and more convenient to use, are not usually adequate to protect against meadow voles.

--------------------- Rabbit ---------------------

Young vines are particularly susceptible to damage by rabbits. Jack rabbits (genus *Lepus*) are the main rabbit pest, although cottontails do cause problems in some areas. They chew or cut young canes near the ground or as high as they can reach.

Description

About the size of house cats with large, long ears, short front legs and long hind legs, jack rabbits normally hop about, but they can outrun most breeds of dogs when frightened. They are actually hares, not rabbits, and as hares they are born fully furred with eyes open and are capable of hopping about immediately. Rabbits, including the cottontail, are born blind, naked and helpless. One other distinction between the two is their nests. Cottontails form underground nests and jack rabbits make depressions in the soil, or form nests beneath brush or other vegetation.

Biology

Jack rabbits breed from early spring to late summer, although breeding may continue where winters are mild. Females may produce more than one brood a year, especially on irrigated land. After a gestation period of about six weeks, a litter (usually three or four) is born. Larger litters are produced in spring. An adult female may produce fourteen or more young per year.

Jack rabbits are active from early evening to early morning throughout the year. Their populations fluctuate and usually reach high levels every five to ten years.

Legal Status

Black-tailed jack rabbits are classified as game mammals by the California Fish and Game Code. When found to be injuring growing crops or other property, they may be taken at any time or in any manner by the owner or tenant of the premises. (If leg-hold traps are used, certain trap designs are prohibited.)

---

*Restricted material; permit required from County Agricultural Commissioner for possession or use.

Photo by Bill Clark

Signs that the meadow vole is active include the girdling of a grapevine, left, and a burrow and runways radiating from it, right.

Photo by David Johnson

The jack rabbit's favorite meal in a vineyard consists of young grapevines.

Natural Control

Natural enemies are seldom numerous enough to provide adequate control. Predators of rabbits include hawks, owls, coyotes, bobcats, foxes and weasels. Dogs and cats also prey on rabbits to some extent.

Monitoring Guidelines

Systematic monitoring procedures and economic injury levels have not been established for rabbits in vineyards. A tour through the vineyard in early morning, late evening or at night to look for rabbits or evidence of feeding will help alert the grower to a potential problem.

Management Guidelines

Rabbit control in vineyards includes natural enemies, shooting, exclusion, repellents, poisoning and habitat modification with choice of methods depending on the nature and urgency of the problem. Rabbit populations should be controlled before a severe problem has developed.

Poisoning. Toxic baits are perhaps the most practical and economical control for large numbers of jack rabbits or where large areas are involved, although results are erratic in some areas. Such baits may be used only to control jack rabbits.

Grains such as rolled barley, squirrel oat groats and whole oats can be used as bait. Root vegetables, fruit and alfalfa hay are used in some locations. (Contact your County Agricultural Commissioner regarding availability of this bait.) Anticoagulant-treated baits may also be used. Before setting out strychnine* bait, it will be necessary to prebait (offer untreated bait) for a time to condition the rabbits to feeding on the bait material.

Prebaiting: Prebaiting (and baiting) is most effective if conducted in late afternoon before rabbits feed. This will also prevent feeding on the bait by non-target diurnal species. Set out untreated bait about 100 yards from the vineyard being damaged. Place a small handful of untreated bait every 10 to 15 steps, preferably on rabbit trails going into the vineyard. It may be necessary to treat the damaged area by placing bait at each point where rabbit trails enter. A good location is where trails meet, as rabbits tend to stop at these junctions.

Increase the amount of untreated bait daily until a slight amount remains at each station after a night's exposure. Bait levels can be reduced or eliminated at locations where there is little or no feeding. Continue for three to five days, or until the majority of the prebait is accepted each night. Under dry conditions grain acceptance may be increased by dampening the prebait with water.

Baiting: In late afternoon, pick up or cover with dirt all prebait and replace it with a heaping tablespoon of grain-type strychnine bait at the same spots where prebait has been accepted. One application should be sufficient to control rabbits.

Anticoagulant baits are placed in covered, self-dispensing feeders or nursery flats. Place one to five pounds of bait in the feeder in areas frequented by rabbits such as runways, resting and feeding areas. It may be necessary to move feeders to new locations to achieve acceptance. Bait should be exposed until all feeding ceases, which may be from one to four weeks.

Poisoned bait should not be placed where livestock or humans, especially children, can pick it up. It is important to note other wildlife in the area so that proper precautions can be taken. To protect non-target wildlife, poisoned bait should be picked up or buried by the second day after application. Dispose of dead rabbits by deep burial or burning.

Fences. Where jack rabbits are a constant threat to young vineyards, exclusion may be the best control. A 36-inch-high fence of woven wire or poultry netting of up to 1 1/2-inch diameter mesh (48-inch roll) makes a rabbit-proof fence. The bottom six inches of wire should be bent outward (facing away from the vineyard) at right angles and then buried six inches deep to prevent rabbits from digging under. If deer-proof fences are constructed, the additional expense of making the enclosure rabbit-tight may be worthwhile.

Vine guards. Except for rabbit-proof fences, individual vine guards, solid or in net or mesh form, offer the most protection. Materials include metal, hardware cloth, plastic, paper, cardboard or other fibrous materials. Cylinders of wire mesh sometimes provide more protection from debarking. These can be made of one-inch mesh poultry netting formed around the vine with ends joined and extending high enough so rabbits standing on hind legs cannot reach foliage or canes. Stakes or wooden spreaders may be used to keep wire in position. If 1/4- or 1/2-inch hardware cloth is used and the wire is set several inches in the soil, the vines are also protected in part from meadow mice.

---

*Restricted material; permit required from County Agricultural Commissioner for possession or use.

Repellents. Chemical repellents may provide temporary relief. These are sprayed or painted on the trunks or foliage according to label instructions. Repeat applications may be required to protect new growth or renew repellency lost through rain or irrigation.

Shooting. Under certain conditions shooting can be an effective control method. Systematic patrolling in early morning and late evening may effectively reduce the population and suppress damage in a localized area.

Habitat modification. Rabbits often invade from adjacent fields and removal of vineyard cover crops and weeds does not have much effect on their populations.

Ground Squirrel

The California ground squirrel, *Spermophilus beecheyi*, is responsible for damage, mostly of a nuisance nature, to grapes. However, because ground squirrels live in colonies, nuisance damage can build up significantly when populations are not controlled.

Description

A medium-sized rodent, about 36 to 51 cm (14 to 20 inches) long (including tail), the California ground squirrel has flecked or mottled fur including a long, somewhat bushy tail. Adults weigh .45 to 1.13 kilograms. Ground squirrels live in colonies of 2 to 20-plus and their underground burrows can form extensive interconnecting systems. When frightened, they seek cover in their burrows.

Biology

These squirrels live in a wide variety of natural habitats but may be particularly dense in areas not disturbed by man, such as along road or ditch banks, fence rows, around buildings and within or bordering agricultural crops. They tend to avoid thick chaparral and dense woods, as well as very moist areas. Much of their time is spent in burrows underground.

Active during the day, they are easily seen from spring to fall, especially in warm, sunny weather. During winter most will hibernate, but some young tend to remain active, especially where winters are not severe. Most adults go into hibernation (estivation) during the hottest parts of the year.

An average litter of seven young is reproduced once yearly in early spring. The young remain in the burrow about six weeks before they go aboveground. Primarily vegetarians, they consume green vegetation, such as grasses and forbs, during early spring. When these dry, they switch to seeds, grains and nuts and also begin storing food. They eat fruits and vegetables as well as bark from vines and trees.

Legal Status

Like pocket gophers, ground squirrels are classified as nongame mammals by the California Fish and Game Code.

Injury

Ground squirrels gnaw vines, particularly young ones, removing bark and often girdling the trunk. They may feed on vines and fruit as well and may also gnaw surface plastic irrigation pipe.

Their burrowing can be very destructive. Burrows average 4 inches in diameter, are 5 to 30 feet or more in length and are 2 1/2 to 4 feet below the ground surface. When digging burrows, squirrels deposit large quantities of soil and rock on the soil surface, making large mounds and burrow openings. These present hazards to machinery and make harvesting difficult. Their frequent burrowing around vines can damage root systems and possibly kill the plant. Their common burrowing beneath buildings and other manmade structures sometimes results in a need for repairs or replacement of structures.

Natural Control

Ground squirrels are generally found in open, clear areas, although they need some protection to survive. Removing debris may make the area less desirable to them unless buildings or other structures provide protection. General area cleanup also makes detection of squirrels and their burrows easier, helps in monitoring the population and improves access to burrows during control operations.

Such predators as coyotes, foxes, badgers and other carnivores may feed upon ground squirrels, but predation rarely, if ever, is a major factor in reducing their populations.

Monitoring Guidelines

Monitor areas where ground squirrels are likely to invade the vineyard such as along ditch or road banks or adjacent to crop or other lands where squirrels are present. This can be accomplished by observing the vineyard during morning hours when they are most active. Ideally, records on the types of control used and their success, as well as on changes in the squirrel population, should be maintained.

Management Guidelines

The presence of ground squirrels requires control. They cannot be excluded by fencing, and no type of habitat modification within a vineyard appears feasible. Furthermore, ground squirrels cannot be repelled by any commercial chemical or physical means. Thus, toxic fumigants, poison baits or traps are the only effective controls.

Trapping. This control is time-consuming and is only practical in small areas or where few squirrels are present. Live-catch traps are effective but disposal of live squirrels presents a problem. Because of their disease potential and general pest status, it is not legal to release them elsewhere.

Several types of kill traps can be used. To be effective, they should be placed on the ground near squirrel burrows or runways and be baited, but not set for several days so that the squirrels become accustomed to them. After they have taken the bait, rebait and set the trap. Place bait well behind the trigger of the trap or tied to it. Good baits are walnuts, almonds, oats, barley or melon rinds.

Box-type trap: A trap that kills quickly can be constructed by modifying a box-type wooden gopher trap. Lengthen the trigger slot with a rat-tail file or pocket knife to permit unhindered trigger swing. This allows the squirrel to pass beneath the swinging loop

of the unset trap. If one trap is used, remove the back of the trap and replace it with hardware cloth, allowing bait to be observed from both ends, but forcing the squirrel to enter from the front.

A dual-assembly trap can be made by placing two modified traps back to back secured to a board. A small strip of hardware cloth connects them and forms the baiting area. Place the bait through the open ends of the trap or through a small door cut in the wire.

A multiple trap box can be constructed by placing modified box traps in the entrances or inside a lidded box. When placed inside, traps are assembled side by side, providing a narrow, rear baiting area formed by a perpendicular baffle. Enclosed traps minimize the chances of accidentally catching pets or poultry.

Conibear: Another trap used successfully is the Conibear which is placed in the burrow opening so that

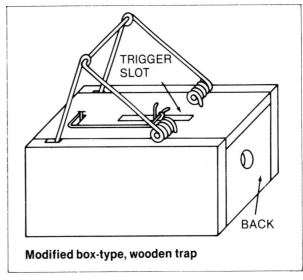

Modified box-type, wooden trap

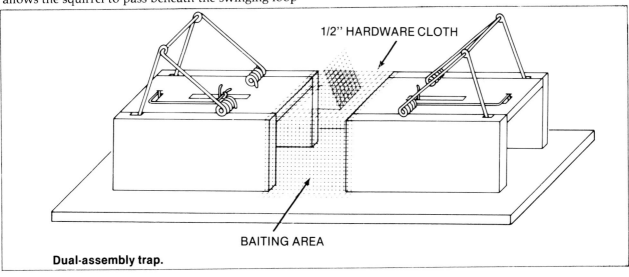

Dual-assembly trap.

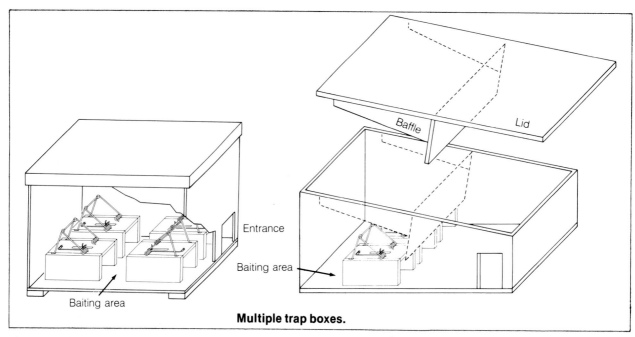

Multiple trap boxes.

the squirrel will pass through it, tripping the trigger. Baiting is usually not required. This trap can also be placed where physical restrictions, such as a hole in the fence, will direct the squirrel through the trap. It should not be placed where pets or other nontarget animals are likely to enter.

Other effective traps for ground squirrels may be available in some areas. With all traps, precautions must be taken to reduce the hazards of trapping nontarget wildlife, pets and poultry.

Fumigation. Most effective in spring or when the soil contains enough moisture to hold gas in a burrow is fumigation outside the vineyard with carbon bisulfide* or methyl bromide.* These materials cannot be used near vines or other desirable plants because contact with roots can injure and kill plants. Gas or smoke cartridges are generally not considered phytotoxic and are frequently used in burrows beneath desirable plants.

Re-treatment is necessary if ground squirrels dig out. Check treatment area after 72 hours and re-treat all opened burrows. Fumigation is not effective during hibernation periods because squirrels plug their burrows with soil.

Carbon bisulfide:* Carbon bisulfide is extremely flammable and very explosive. Use waste balls soaked in carbon bisulfide or a specially designed applicator (Demon rodent gun) to fumigate each active burrow. After application, seal burrow with soil, tamping lightly.

Methyl bromide:* Methyl bromide is applied to each active burrow with a specially designed nozzle. Place in the burrow, plug the entrance with soil and dispense the fumigant. Withdraw hose and seal burrow.

Gas cartridge: The U.S. Fish and Wildlife Service produces a relatively safe and easy-to-use gas cartridge designed for fumigating burrowing rodents. A mixture of chemicals, the cartridge gives off smoke and other toxic gases when ignited; use one or two when an active burrow is observed. With a nail or sharp object about the diameter of a pencil, puncture the cap end of the cartridge, and rotate the nail to loosen the materials inside. Insert the fuse in the center hole. Place the cartridge as far back as possible in the burrow and light the fuse. With a shovel handle or stick, push the cartridge down the burrow and quickly seal the opening with soil, tamping lightly. Seal connected burrows where smoke is seen escaping. Larger burrow systems usually require two or more cartridges, which can be placed in the same burrow or in connecting burrow openings.

Smoke coming from the cartridge occasionally ignites, presenting danger of fire. Fumigants should not be used beneath buildings occupied by humans or desirable animals.

Most County Agricultural Commissioners sell gas cartridges. Several commercial gas cartridges are also available at retail outlets.

Toxic baits. These have been developed for controlling ground squirrels; they accept grain bait only when

---

* Restricted material; permit required from County Agricultural Commissioner for possession or use.

the green vegetation of winter and early spring is gone. In late spring they usually eat grain bait, but when hot weather begins a large percentage go into summer hibernation. In fall most squirrels are active and will feed on grain baits until their winter hibernation. Various grain bait formulations are available from commercial distributors or the County Agricultural Commissioner.

Single-dose baits: Toxic baits containing zinc phosphide* or strychnine* are effective for controlling ground squirrels in many situations. Bait distribution may be done by spot baiting—placing bait by hand on bare ground to cover two to three square feet at the side or behind each active burrow, or by broadcasting by hand or machine—scattering bait over an entire infested area. Consult product label for proper application methods and rates.

Anticoagulants: Toxic baits containing anticoagulants cause death by interfering with the animal's blood clotting mechanism. They are effective only when consumed in several feedings for five or more days; effectiveness is greatly reduced, if 48 hours or more lapse between feedings. This, as well as the availability of an effective antidote, makes these materials relatively safe to livestock, pets and children.

Anticoagulant baits can be used (1) in bait boxes or (2) by repeated spot baiting. Bait boxes are small structures that the squirrel must enter to eat the bait and are preferred around homes because they prevent children and pets from reaching the bait.

*Restricted material; permit required from County Agricultural Commissioner for possession or use.

Bait box baiting: Bait box entrance hole(s) should be 2 1/2 to 4 inches across to allow entry by squirrels but not by larger animals. A lip or some other arrangement should be used to prevent bait from falling from the box when squirrels move in and out. The box should be lockable or otherwise difficult for children to open and should be secured so that it cannot be turned over or easily removed. A self-feeding arrangement should be incorporated to allow a continuous supply of bait.

Place bait boxes containing one to five pounds of bait in areas frequented by ground squirrels. Inspect bait stations daily and add fresh bait as needed; replace moldy or old bait. Squirrels may require a number of days to get accustomed to the bait box. Anticoagulant baiting using a bait box generally requires two to four weeks or more to be fully effective. Continue baiting until all feeding ceases and no squirrels are observed. Baits should be picked up and disposed of upon completion of the control program.

Eating anticoagulant bait does not immediately affect the squirrels' feeding or activity, and they will appear healthy while feeding at the bait boxes. They will soon be affected (usually after three days of feeding), but it is extremely important that bait be continually available.

Spot baiting: Repeated spot baiting (not using a bait box) with anticoagulant bait can effectively control squirrels, but it is only permissible if recommended by the product label. Scatter a handful of bait (about 10

Uncontrolled, the California ground squirrel can cause significant damage to young vines.

placements per pound) evenly over 40 to 50 square feet near active burrows or runways. Re-treat every other day for three or four times so an uninterrupted supply of bait is available for six to eight days. Scattering of bait takes advantage of the squirrel's natural foraging habits and helps prevent domestic livestock and wildlife from picking it up. Never put squirrel bait in piles. Also, because ground squirrels do not feed in their burrows, bait should not be placed there.

Anticoagulants have the same effect on nearly all warm-blooded animals, including some birds. Cereal baits are attractive to some dogs as well as other non-target animals, so care must be taken to prevent their access to the bait. Danger to children and pets can be greatly reduced by using, as mentioned, a bait box or otherwise placing the bait out of their reach. In the case of accidental ingestion by humans or pests, contact a physician or veterinarian immediately.

Deer

Where the habitat is favorable, deer can significantly damage vineyards. In areas of high deer population, this is a major obstacle to successful viticulture. Foothill and coastal districts with woodlands that provide deer cover usually experience the heaviest depredation. Some valley vineyards near wooded areas or stream bottoms may also suffer damage. State game management laws limit control methods available to growers.

Description

Deer are large gray or brown animals difficult to observe because they are night feeders. Hoofprints in the vineyard indicate their presence. Deer hooves are split and about two to three inches long; they are pointed at the front and more rounded at the rear, unlike those of pigs and sheep, which are rounder in front. Tracks of running deer show a spreading of the hoof split.

Biology

Deer usually seek shelter in adjacent wooded or brush areas but they may live within the vineyard. Some are permanent residents of the area; others are migratory and spend winter and spring around the vineyard and summer and fall at higher elevations.

Legal Status

Deer are classified as game mammals by the California Fish and Game Code. When found damaging growing crops, a permit may be obtained from local game wardens to control them by shooting. Growers must adhere to permit regulations, including field dressing and delivery of animal carcasses as designated by the game warden. Other methods of destroying deer are illegal.

Injury

Deer may almost completely strip vines of foliage. Severe stunting of vines can result from repetitive deer browsing. Young vines may be damaged by buck deer rubbing their antlers on the trunks, arms, or cordon branches. This usually causes severe breakage or scarring.

Management Guidelines

Fencing. Fencing is the most effective method of excluding deer from a vineyard. A six-foot woven mesh wire fence is the minimum recommended height for level ground. A seven-foot fence may be necessary in some areas. Deer being chased may clear an eight-foot fence or if the fence is on sloping terrain. A five- or six-foot-high mesh fence can be heightened with two or three strands of barbed wire.

Fences are sometimes constructed only along the side of the vineyard adjacent to uncultivated land where deer are usually found. While this is less expensive, it is rarely satisfactory because deer will frequently go around the ends of the fence.

Deer fences must be checked periodically. Damaged wire, broken gates, soil washouts beneath fences, etc., permit access and must be repaired. This job becomes increasingly difficult as fences age and become more vulnerable to breakage. Care must be taken not to fence deer in the vineyard.

Repellents. Repellents are used to prevent damage to vines. Many odor repellents have been tried, but deer usually adjust to them, especially when hungry. When deer populations increase, severe competition for food results and repellents become much less effective.

Taste repellents, especially products containing Thiram, can be effective if applied to new foliage as it develops. This requires spraying every one or two weeks as new foliage comes on. A small backpack sprayer can be used to spray each new, unsprayed shoot tip. Thiram tends to wash off from rain or sprinkler irrigation.

Such noisemaking devices as carbide guns and electronic alarms have not generally been effective in repelling deer from a vineyard since deer rapidly adjust to these scaring devices.

Hunting. In California permits can be obtained to kill deer when they are involved in crop depredation. This is not a long-term solution, but it does offer relief when deer are numerous. Shooting may be the only practical means of removing a deer trapped in a fenced vineyard.

Information and depredation permits may be requested from your local game warden.

Habitat management. Eliminating suitable shelter for bedding and other survival needs of deer is rarely possible. Grapes are a favored food, especially when new foliage is developing, and planting other food near the vineyard will probably not prevent damage. This practice may result in a general increase in the area's deer herd.

BIRDS

Birds are common pests in backyard vineyards as well as in large vineyards. Their damage is restricted to fruit and monitoring for potential damage should be initiated when the fruit begins to ripen and change color. Early detection will greatly increase management success.

As with rodents, clean cultural practices may contribute to fewer bird problems. Many birds, such as the house finch, sparrow and dove, are attracted to vineyards by the available nest and loafing habitats in the form of weedy ditches, hedge rows and brush or trash piles. Seed-eating birds entering a vineyard to feed on weed seeds, may later turn to ripening grapes.

As in rodent control, several factors must be considered: (1) Species identification is critical. Different species will require different management techniques. (2) If birds are drawn to the vineyard by the surrounding habitat, alter the habitat. (3) Initiate a management program compatible with the agro-ecosystem. (4) Continue to monitor bird populations and if necessary alter management as conditions change. Integration of more than one technique generally proves effective.

Bird control materials are not as readily available as those for rodent control. Certain materials are available from some County Agricultural Commissioners. Local county farm advisors and Agricultural Commissioners can assist in locating control measures.

House Finch (Linnet)

The house finch (linnet), *Carpodacus mexicanus*, is one of the most serious pest birds in California vineyards. Vineyards and the habitat around them are usually attractive to it so few vineyards escape at least some damage.

Description

The house finch, about the size of a house sparrow, has a heavy bill. Adult males are brownish with a bright red breast and forehead. Careful observation will reveal a red stripe over the eye and rump. Females are gray-brown above with dusky colored streaks on their underparts.

Biology

Primarily a seed-eating bird, the house finch can be found throughout California. The female builds a shallow nest of any material available in a variety of sheltered places. Nesting begins as early as March in warm areas but may be delayed as late as July in colder areas. Two broods are often raised from the same nest.

Pairs are well scattered during the nesting period. As the young are fledged, they form local feeding flocks, often joined by a few adults. By mid-August these flocks comprise most of the young birds of the area.

Photo by J. P. Clark

Evidence of what kind of damage deer can do to vines is seen in the stripped cane.

Photo by Bill Clark

Grapes in center photo were pecked by house finches. A female finch is at left, a male is at right.

GLOSSARY

Vertebrate Pest Control

Acute rodenticide—Toxic compound specifically for rodents that produces death from a single feeding.

Avicide—Toxic compound specifically for birds.

Bait box—Small structure in which bait is placed so that the animal must enter to take the bait.

Broadcast baiting—To scatter bait over an infested area. Usually done by hand, machine or aircraft.

Browsing—To feed on leaves, shoots, etc. of plants and trees.

Communal roost—Roost used in common by a large number of birds.

Control—To regulate, restrain or curb the population of a vertebrate species that has become a pest.

Cultural control—To make a habitat unsuitable for a pest by changing a cultural practice.

Debarking—Removal of bark from a tree or woody plant by an animal.

Depredation—When a pest animal or species despoils or "plunders" a crop.

Diurnal—Animal or species of animal active during daylight hours.

Estivation—State of dormancy occurring during summer months.

Exclusion—To prevent one or more animals or species of animals from entering an area.

Feeding flock—Flock of birds banded together for the purpose of feeding.

Flyway—Aerial path habitually used by a species or species' of birds to travel from one point to another.

Frightening device (gas cannons, shell crackers, flags, noisemakers, etc.)—Any device used to drive a pest species away from a crop or an area by scaring it.

Fumigant—Substance that produces toxic or suffocating gases.

Game mammal—Mammal specified by the California Fish and Game Code to be hunted for food or sport.

Gestation period—Period from conception to birth.

Girdle—Gnawing cut made by an animal through and encircling the trunk or limb bark of a tree or other woody plant.

Habitat modification—To alter the environment where an animal species is found. In pest control this is sometimes done to make the habitat less favorable for the species concerned.

Hibernation—State of dormancy occurring during winter months.

Introduced—Species established as a new element in an environment.

Leg-hold traps—Device for trapping and holding or restraining an animal by the leg.

Lethal dose—Quantity of a toxicant necessary to cause death.

Loafing area—Area where birds loiter during daylight hours.

Management—Skillful regulation and control of an animal species and/or its environment.

Mechanical baiting—Use of a device that automatically dispenses bait.

Migratory—Making an annual, regular round trip between two geographic regions.

Multiple-dose bait—Poisonous bait that requires a sustained dosage over a period of time to produce death. An example is an anticoagulant.

Native bird—Bird that is an original species in an area or region. Indigenous.

Natural control—Regulation or restraint of an animal population in a manner conforming to nature. Examples: predators, diseases, lack of food.

Netting—Nets used to protect a crop from birds.

Nongame mammal—Any animal not commonly hunted, as specified in the Fish and Game Code.

Nontarget species—Any species that is not the object of the controls being applied.

Phytotoxic—Injurious and sometimes lethal to plants.

Prebaiting—Placing of nontoxic bait to condition a pest species to eating it before toxic bait is applied.

Predator—Any vertebrate that survives by the regular taking of another vertebrate (or insect) for food.

Resident—Species that lives in an area year round. Does not migrate.

Runway—Path that an animal or animals commonly travel over or through.

Single-dose bait—Toxic bait that produces death from one dose. Also called acute toxic bait.

Spot baiting—Placing of bait by hand at selected sites.

Untreated bait—Bait to which no toxic or repellent substance has been applied. Used for prebaiting.

Vertebrate pest—Any species of vertebrate animal, in any area, that becomes a health hazard, causes economic damage or is a general nuisance to one or more persons.

(a)

(aa)

(bb)

(b)

(c)

(cc)

(dd)

(d)

On the preceding page:

The four perennial weeds characterized as the most "troublesome" in vineyards are (a) johnsongrass, (b) bermudagrass, (c) field bindweed and (d) yellow nutsedge (there is also a troublesome purple nutsedge). In the smaller photos are (aa) the collar region of a johnsongrass blade, (bb) a stalk of bermudagrass, (cc) the blossoms of field bindweed and (dd) the nutlets or tubers of yellow nutsedge.

Section XIII—WEEDS

Contents

This section was prepared by Joyce K. McReynolds, Editor, Agricultural Sciences Publications, Berkeley, with the assistance of Curtis D. Lynn, County Director, Cooperative Extension, Tulare County, from information compiled by the following weed scientists: Arthur H. Lange, Weed Scientist, San Joaquin Valley Agricultural Research and Extension Center, Parlier; Bill B. Fischer, Weed Farm Advisor, Fresno County, and Harold M. Kempen, Weed Farm Advisor, Kern County.

The following also provided information for this section in interviews on local weed problems: Keith W. Bowers, Farm Advisor, Napa County; L. Peter Christensen, Farm Advisor, Fresno County; Frederik L. Jensen, Viticulturist, San Joaquin Valley Agricultural Research and Extension Center, Parlier; Amand N. Kasimatis, Viticulturist, Cooperative Extension, Davis; Donald A. Luvisi, Farm Advisor, Kern County; Rudy A. Neja, Farm Advisor, Monterey County, and William L. Peacock, Farm Advisor, Tulare County.

Weeds compete with grapevines for water, soil nutrients, and sometimes light, and most will outgrow newly planted vines. Once vines have been trained, they can compete more effectively with annual weeds, but their growth and yield can be greatly reduced by heavy stands of annual or perennial weeds, and vineyards with dense weed growth require more water and nitrogen fertilizer.

Some weeds may interfere with cultural and harvest operations; others may inhibit control of insect pests. Also, some weeds produce seeds that attract birds which, once they become habituated, can cause much crop damage. Other problems include difficulties in controlling gophers and field mice and the fact that dry weeds create a fire hazard.

By definition, a weed is a plant that is growing out of place. It is undesirable; it is generally combatted. But in a vineyard weed management program, there are times when weeds are encouraged to grow, and here we enter into an area of differences in terminology that are arguable among the experts: when you're letting it grow, it is called "natural vegetation" or "grasses"; when you're fighting it, it is a weed. In the following discussion, the word "weed" can be interpreted as friend or foe, depending on context. Occasionally, the words "natural vegetation", "grass(es)", or "plant(s)" are employed, either for the sake of variety or for the sake of explicitness.

Bibliography

Growers Weed Identification Handbook, Priced Publication 4030, Division of Agricultural Sciences, University of California, Berkeley, California. For sale by Agricultural Sciences Publications, University of California, Berkeley, CA 94720.

Weeds of California. Robbins, W. W.; Bellue, Margaret K., and Walter S. Ball. State of California Department of Food and Agriculture, first printing 1951; second printing 1970. For sale by Documents and Publications, P.O. Box 20191, Sacramento, CA 95820.

CHARACTERISTICS OF IMPORTANT VINEYARD WEEDS

Numerous publications describe the botanical characteristics of common California weeds. Growers may find *Weeds of California* and *Growers Weed Identification Handbook* useful in identifying weeds in their vineyards. (See bibliography this page.) Knowledge about growth habits and methods of reproduction and spread is essential for effective, economical weed control.

From the standpoint of the grower, weeds fall into two broad categories, annuals and perennials, and are distinguished by their reproduction methods. Annuals usually reproduce from seeds and complete their life cycle in one growing season, although crabgrass, a common annual grass in vineyards, can also spread from stolons. Control involves interrupting this cycle by killing the plant before it can produce and scatter its own seeds. However, many seeds with hard outer shells remain viable for several years, making control difficult.

Perennials also reproduce from seeds, but they may produce underground stems (rhizomes) or tubers, or aboveground runners (stolons), which can send out roots and shoots and produce new plants that are viable even when detached from the parent plant.

Annual Weeds

Many annual weeds are found in California vineyards. For the most part their influence on growth and yield is not as important as that of perennial weeds. They are usually spread by seed, and seed production varies with species. Many grasses produce 10,000 to 30,000 seeds per plant, while some broadleaf weeds produce more than 250,000 seeds per plant. Very little data have been collected on the number of seeds produced by different species, but generally the soil is a reservoir for 40 million or more seeds per acre, if little effort is made to reduce their numbers.

Certain seeds are rapidly spread by wind. Examples: dandelion, sowthistle, prickly lettuce, mare's tail, flax-leaved fleabane, groundsel and cudweed. They can blow in from adjacent fields, ditches and roadsides and are therefore difficult to keep out of vineyards. Others are spread by mechanical equipment. Seeds of nightshade, pigweed, Russian thistle and annual morning glory may be spread by mechanical harvesters; puncture vine and sandbur can be spread by vehicle tires and shoes.

Nightshade family weeds are poisonous, their fruit conceivably a problem, if harvested with grapes. Others such as sand bur, stinging nettle, green amaranth, mare's tail and lambsquarters are bothersome to hand crews; pigweed (*Amaranthus* spp.), turkey mullein and many other weeds can cause skin irritations and allergic reactions.

Many annual weeds may host viruses and other diseases, nematodes and insect pests. Many broadleaf species are hosts of root knot nematode. London rocket and other mustards may contribute to a buildup of false chinch bugs, which can move onto grape shoots when the weeds mature and dry in the spring. Other weeds harbor thrips.

Perennial Weeds

Numerous perennial weeds can infest vineyards, but the most troublesome are johnsongrass, bermudagrass, field bindweed, and yellow and purple nutsedge. Less frequently occurring perennials include whitehorsenettle, Russian knapweed, mullein, asparagus, dallisgrass and the biennial cheeseweed, which behaves as a short-lived perennial.

Johnsongrass (*Sorghum halepense* (L.) Pers.) is a pernicious tropical plant that reproduces from overwintering rhizomes and seeds, both produced in quantity. Foliage is vigorous and tall (usually 3 to 5 feet with maximum growth up to 9 feet). Although winter frost kills some underground rhizomes, they will overwinter where soils do not freeze to depths of 18 to 24 inches. Axillary buds every inch or so along the rhizomes are independently capable of developing into new shoots the next season. One plant can develop 5,000 buds in one season. Such shoots are more vigorous than those produced from seed. Cultivation can encourage more buds to sprout and spread if rhizomes are cut up into small sections and scattered.

Seed production is the primary source of dissemination. Seeds travel in water and in the manure of birds and animals; they are also carried by wind, animals and machines. They germinate readily in shallow soil when moisture and light are present, but deep tillage may preclude germination of some seeds for as long

as seven years. Natural dormancy is common during the year they are produced.

Because johnsongrass competes aggressively in vineyards, a preventive program is essential and all plants must be eradicated, including those on the vineyard's peripheries. A communitywide eradication and control program (such as one organized in the Imperial Valley) would be logical against this weed.

Bermudagrass (*Cynodon dactylon* (L.) Pers.) is a serious weedy grass pest in vineyards. A long-lived perennial, it reproduces primarily by stolons on the surface of most soil types, but prefers those that are moist, warm and well drained. It also reproduces from seed, although seed production is sparse. Bermudagrass is intolerant of dense shade so that it does poorly in the vine rows of varieties which grow luxuriantly and are trained to wide trellises. It is also intolerant of desiccation and frost.

Bermudagrass, a host for several nematodes (*Meloidogyne* spp., *Pratylenchus* spp.), is also a host for Pierce's disease, which can be devastating to grapevines.

For both johnsongrass and bermudagrass, prevention is the best practice; once established they are difficult to control (see *Special Weed Problems*).

Field bindweed (*Convolvulus arvensis* L.), a perennial of the morning-glory family, is most noxious on (but not limited to) heavier soils along California's coast and nonirrigated valleys. Its creeping, twining stems enable it to grow up the vine and stake and over grape foliage, severely competing for light. Flowers are funnel-shaped, white or pinkish, sometimes with reddish or purple stripes. Another morning glory, *Ipomea* spp., an annual with larger white, red, purple or blue-purple flowers that are frequently variegated, can be mistaken for it.

Bindweed's root system is extensive, as deep as 10 feet, extracting soil moisture to the extent that it can severely reduce grape production in nonirrigated areas. It is spread by abundantly produced rhizomes, but seeds are also a principal source of reproduction. They have hard seed coats, remaining viable in the stomachs of birds and animals for 10 days or more and in soils for as long as 40 years. Soil fumigants are not effective.

Combinations of tillage and herbicides are the most logical programs, but are not fully satisfactory against this serious weed pest.

Yellow nutsedge (*Cyperus esculentus* L.) and **purple nutsedge** (*C. rotundus* L.) are sedges that are not severe competitors in vineyards, except during the vineyard's first few years. However, they can become dense in vineyards where other weeds are controlled with herbicides, competing for moisture and nutrients. Essentially aquatic weeds, they are especially troublesome around the emitters in drip-irrigated vineyards and are frequently a problem in furrow-irrigated vineyards on heavier, shallow soils where the water "subs" substantially.

Both sedges reproduce and spread primarily by tubers. Yellow nutsedge also produces quantities of viable seed; purple nutsedge, on the other hand, produces few viable seeds with short-term dormancy. Tubers normally remain viable in moist soils for up to two years and are capable of producing up to six shoots from axillary buds. Purple nutsedge tubers are produced in chains on rhizomes and are bitter-tasting, dark brown and hairy. Their growing tips are dormant as long as they remain attached to the chain. The tubers of yellow nutsedge are roundish and are borne singly (not in chains).

Nutsedge is a host for root knot (*Meliodogyne* spp.) nematode.

Both nutsedges are relatively susceptible to shading, with reductions in the number of tubers per plant and the amount of foliage directly correlated to amount of shading. Purple nutsedge tubers can be desiccated in hot summer weather if soils can be repeatedly tilled and thoroughly dried. Eradication in established vineyards is extremely difficult and may not be economical since competition with established vines is not severe.

Listed with the following summer annual and perennial weeds often found in vineyards are subjective ratings of their competitive ability and the problems they present at harvest. Also included is a listing of commonly found winter annuals.

Page No.[1]	Summer Weeds in Established Vineyards	Competition ranking[2]	Harvest problem[2]
13	Amaranth, green or careless weed (*Amaranthus hybridus*)	9	9
33	Barnyardgrass (*Echinochloa crusgalli*)	7	3
19	Cocklebur (*Xanthium strumarium* L. var. *canadense*)	7	9
32	Crabgrass (*Digitaria sanguinalis*)	5	2
42	Cupgrass, southwestern (*Eriochloa gracilis*)	2	3
38	Feather fingergrass (*Chloris virgata*)	2	3
77	Fleabane, flax-leaved (*Conyza bonariensis*)	8	8
84	Foxtail or bristlegrass (*Setaria* spp.)	2	2
157	Groundcherry (*Physalis* spp.)	3	3
70, 86, 167	Lovegrass (*Eragrostis* spp.)	2	2
16	Mare's tail (*Conyza canadensis*)	8	10
112, 166	Morning glory (*Ipomoea* spp.)	9	9
9, 142	Nightshade (*Solanum* spp.)	7	7
57	Pigweed, prostrate (*Amaranthus blitoides*)	3	2
53	Pigweed, redroot (*Amaranthus retroflexus*)	8	8
79	Pigweed, tumble (*Amaranthus albus*)	5	6
24	Puncture vine (*Tribulus terrestris*)	3	9[3]
27	Purslane (*Portulaca oleracea*)	2	2
21	Russian thistle (*Salsola iberica*)	5	7
40	Sandbur, longspine (*Cenchrus longispinus*)	3	9[3]
39	Sprangletop (*Leptochloa* spp.)	3	3
28	Spurge (*Euphorbia maculata*)	2	2
11	Sunflower (*Helianthus annuus*)	7	7
55	Turkey mullein (*Eremocarpus setigerus*)	2	2

[1]Page number in *Growers Weed Identification Handbook*.
[2]Ranked 0 to 10: 0 = no effect; 10 = severe.
[3]For table and raisin grapes.

Page No.[1]	Perennial Weeds in Vineyards	Competition ranking[2]	Harvest problem[2]
72	Bermudagrass (Cynodon dactylon)	8	5
	Bracken fern (Pteridium aquilinum var. subescens)	5	5
51	Curly dock (Rumex crispus)	3	2
41	Dallisgrass (Paspalum dilatatum)	6	5
95	Dandelion (Taraxacum officinale)	3	2
74	Field bindweed (Convolvulus arvensis)	9	7
85	Foxtail, knotroot (Setaria geniculata)	3	3
71	Johnsongrass (Sorghum halepense)	9	9
73, 82	Nutsedge (Cyperus spp.)	3	2
60	Silverleaf nightshade (Solanum elaeagnifolium)	7	5

[1]Page number in Growers Weed Identification Handbook.
[2]Ranked 0 to 10: 0 = no effect; 10 = severe.

Page No.[1]	Winter Weeds	Maximum growth (feet)
65	Annual bluegrass (Poa annua)	1
67	Bromegrass, ripgut (Bromus spp.)	2
20	Cheeseweed, malva (Malva parviflora)	8
50	Chickweed (Stellaria spp.)	2
25	Cudweed (Gnaphalium spp.)	1
6	Fiddleneck (Amsinckia intermedia)	4
111	Filaree, redstem (Erodium cicutarium)	2
77	Fleabane, flax-leaved (Conyza bonariensis)	2
3	Groundsel (Senecio vulgaris)	1
17	Henbit (Lamium amplexicaule)	1
16	Horseweed, mare's tail (Conyza canadensis)	7
12	Knotweed (Polygonum aviculare)	1
22	Lambsquarters (Chenopodium album)	7
5	London rocket (Sisymbrium irio)	4
54	Minerslettuce (Montia perfoliata)	1
8	Mustard (Brassica spp.)	7
46	Nettle (Urtica urens)	2
1	Prickly lettuce (Lactuca serriola)	6
10	Shepherd's purse (Capsella bursa-pastoris)	2
4	Sowthistle (Sonchus oleraceus)	6

[1]Page number in Growers Weed Identification Handbook.

SOME BENEFITS OF WEED COVER

Historically, weed control in nonirrigated vineyards has meant elimination of weeds at the end of the winter rainy season and subsequent control by cultivation so that weeds would not take moisture from the soil during the dry summer. Before the development of herbicides, so-called "clean culture" was practiced in square-planted vineyards using cross-cultivation and hand hoeing or shovelling around the vine trunks. Hand operated plows were effective, but they required extra labor to guide the plow as it moved in and out of the vine row.

Clean culture is still the ideal of many growers, particularly in nonirrigated north coast vineyards and in San Joaquin Valley raisin vineyards, but it can be difficult and time consuming. With the current practice of staking and wire trellising in rows and the increasing practice of irrigating vineyards, some weed growth can be tolerated, and indeed, has been found to have its uses.

Erosion control is a management problem in several California grape growing regions. On northern hillside vineyards and in coastal counties subject to heavy fall and winter rains, native vegetation may be allowed to grow during winter, particularly in young vineyards. The vegetation's fibrous root system binds the soil particles. Sometimes vineyards are seeded just before or immediately following harvest with such cover crops as barley, annual rye, brome, etc. The native vegetation or cover crop is then disked into the soil in early spring to provide an organic layer which prevents erosion from spring rains.

Similarly, in some windy, sandy-soil regions, particularly in San Bernardino County, weeds or cover crops are allowed to grow to prevent wind erosion. Here the common practice, adopted from the Midwest, is to use a graham plow which cuts roots but leaves the plant standing upright and its erosion control properties intact.

Water penetration, another problem, can sometimes be partially solved by allowing some weed growth. Weed roots penetrate the soil, providing natural channels for water when they die and decay. The weeds themselves facilitate penetration by slowing down surface water movement.

Cover crops can be beneficial. They provide erosion control and facilitate water penetration, as mentioned previously. They also add humus to the soil, and such legume cover crops as clover species, bell beans, etc. may also add nitrogen to the soil. In some areas, a 3- to 4-foot band under the vine row is kept weed-free, and a cover crop is sown, or naturally occurring plants encouraged in the row middles.

Dust suppression is important in table grape vineyards, and some wine grape growers favor it, too. Annual grass culture, where weeds are mowed every few weeks in the growing season, reduces dust, not only by covering the soil, but by eliminating the need for cultivation during the growing season's driest part. Many growers feel that suppression of dust is a factor in controlling spider mites.

Higher humidity and heat reduction are two other benefits that can be gained from the practice of allowing naturally occurring vegetation to grow in summer, some growers believe. A humid environment is considered favorable for spider mite control, but may increase powdery mildew. Lowering vineyard temperatures may lessen the risk to the berries of heat burn when early season temperatures exceed 38°C (100°F).

Table grape quality improvements have also been attributed to grass culture in the San Joaquin Valley. While vines have high nutrient requirements early in the season, after July the presence of high levels of nitrogen in soil may retard fruit color in colored varieties of table grapes. It is thought, therefore, that weed competition for nutrients and a corresponding slowing in vine growth rate later in the season help produce high quality table grapes in some varieties. This may, however, reduce total yield as has been demonstrated experimentally.

VEGETATION MANAGEMENT GUIDELINES

With the advent of vineyard irrigation, the concept of weed control has largely changed from one of eradication to one of "weed management," of holding weed growth in check so that the weeds do not compete so severely for water or nutrients or interfere with pest control or harvesting. A management program must consider weed control in both the vine row and the row middles. Components of the program vary according to whether the vineyard is nonirrigated or irrigated by furrow, sprinkler or drip methods and whether a grower is producing grapes for wine, table or raisins. Other factors might include climatic conditions, topography, soil type, grape variety, prevalent weed species and the cultural methods preferred by individual growers.

Bearing in mind the above, the following are broad, general examples of weed control programs.

Established Vineyards

If a grower plans to use strip chemical weed control, residual chemicals such as simazine (Princep), diuron (Karmex), napropamide (Devrinol), etc., are typically sprayed in a 4-foot-wide band down the vine row only. Spraying is done in winter or spring, as soon as equipment can get into the vineyard without compacting wet soil and generally after vines have been pruned and tied. To be effective, application of residual chemicals must be followed within a few weeks and sometimes sooner, by rainfall or sprinkler irrigation, depending on the herbicide used.

If cultural methods are preferred to chemical control, a row plow (sometimes called a French plow or a kirpy plow) is used in the trellised vine row. A tripping mechanism on the plow is activated by a lever contacting the vine trunk and stake, causing the plow to move around the vine. When properly adjusted, it will throw the mat of weed growth towards the row middles. Weeds are thus removed from the row, except

1/2 inch or more rainfall does not occur within 10 days, sprinkler or flood irrigation is necessary.

Oryzalin (Surflan) is registered for use in vineyards. It is very insoluble in water and has to be incorporated by rainfall or sprinkler irrigation soon after application. However, in cool seasons and in some soil types, irrigation has been delayed as long as one month without loss of activity. Oryzalin can be used safely on all soil types and is very low in toxicity to animals.

Application and rates. Rates of 2 to 4 pounds of active ingredient per acre can be used in vineyards. The higher rate will control many broadleaved weeds as well as annual grasses.

Postemergence (Foliar-Applied) Herbicides

Contact herbicides such as weed oil, dinoseb* formulations and paraquat* can be used to control those weeds not controlled with the preemergence herbicides.

A spray made by mixing 1 to 2 quarts of dinoseb (5 pounds of active ingredient per acre formulations), 10 to 20 gallons of weed oil, and 80 to 90 gallons of water to make up 100 gallons of spray is very effective for the control of established weeds. Care must be exercised not to spray the young vines. A wetting agent at ½ percent of the spray volume (one-half gallon per 100 gallons) can be substituted for the weed oil if the weeds are young.

SPECIAL WEED PROBLEMS

Johnsongrass and Bermudagrass

Control of both these weeds is difficult. Mechanical control is usually used in the row middles, but implements such as harrows may spread these pests. Chemical control, requiring accurate timing and application, is necessary in the vine row. Slanting knives or sweeps that run below the soil surface may be used to supplement chemical control in the vine row.

In older vineyards, one successful technique is to allow johnsongrass and bermudagrass to grow early in the spring, and then to direct glyphosate (Roundup) sprays onto the weeds below the vine shoots just before the shoots begin to bend or trail downward. This permits maximum growth of these

perennials so that they are moving some metabolites into their root systems and the glyphosate will therefore be translocated into the entire plant. Residual herbicides, such as napropamide, trifluralin, oryzalin, diuron, or simazine will control johnsongrass seedlings.

A second technique utilizes trifluralin at maximum label rates. The vine row is row plowed when the vines are dormant, moving the stolons and rhizomes toward the row middles. In heavily infested vineyards row plowing two times in opposite directions may be required. Any remaining stolons and rhizomes are hand hoed from around each vine trunk. The soil and weeds are disked well to cut up all the stolons and rhizomes into short segments, and trifluralin is applied at maximum rates and disked in. Then the treated soil is thrown back into the vine row. Subsequent spot treatment of weed regrowth with glyphosate or dalapon may hasten eradication.

This technique can provide more than 75 to 90 percent control the first season and if repeated the second season, eradication can be achieved. Followup with residual herbicides will be required for five years or so to eliminate reestablishment from seeds present in the soil.

In the past, geese have been used in established vineyards heavily infested with johnsongrass. Vineyards must be fenced to keep the geese in and dogs out, water must be provided, and the geese must be removed before the berries begin to ripen. These factors make weeder geese impractical for routine weed control, but their predilection for johnsongrass (they will not eat many other weeds) has permitted some growers to completely eradicate this weed pest where other methods have failed.

Field Bindweed (Perennial Morning Glory)

Very frequent disking of row middles helps to keep bindweed temporarily under control. However, disking has no effect on bindweed in the vine rows. Row plowing, weed cutting and repeated applications of weed oil have resulted in only temporary and expensive control and these practices can injure young grapevines. Contact herbicides such as paraquat or dinoseb are much safer for vines than oil if kept off the foliage and can be sprayed repeatedly in the same manner as weed oil, although such sprays are expensive. One-half to 1 pound of paraquat per acre with a surfactant is sufficient to burn back bindweed in the vine row if it is thoroughly wetted. The results are

---

*Restricted material; permit required from County Agricultural Commissioner for possession and use.

only temporary, however, and the bindweed will re-establish itself as soon as the grower relaxes control.

An effective way to control bindweed is to apply trifluralin by subsurface layering. The herbicide is applied by means of a spray blade or in a row plow furrow at a soil depth of 4 to 6 inches. The growing tips of the bindweed shoots cannot penetrate or are destroyed when they contact the thin, concentrated layer of chemical. Dormant-season application offers the advantage of timely use of labor, and appears to work better than late spring or summer applications. A weakness of this method is the tendency of bindweed to flourish at the edge of the treated area and adjacent to the stakes and trunks of vines and in soils that develop deep cracks upon drying.

The use of layered trifluralin has shown no indication of phytotoxicity to bearing vines, and annual weed control has been achieved in addition to superior bindweed control. However, weed species such as those in the mustard family, including shepherd's purse and London rocket and weeds in the nightshade and thistle families, are somewhat tolerant of layered trifluralin, necessitating the use of contact herbicides.

Winter applications of oryzalin, simazine or diuron will control field bindweed seedlings where these can be safely used.

Glyphosate has been somewhat erratic in control of bindweed in vineyards. Effective control has been obtained when glyphosate was applied on vigorously growing field bindweed. However, when vine shoots are trailing the ground, glyphosate treatments are considered hazardous. Only slight spray drift or contact on one or two grapevine leaves will cause foliar injury the following spring.

A Note on Herbicides Registered for Use in 1981

Since this publication was first printed in May 1981, Oryzalin (Surflan) has been registered for use in bearing as well as nonbearing vineyards and Oxyfluorfen (Goal), previously omitted in this publication, is registered for use in bearing vineyards only.

Oryzalin, a relatively broad spectrum herbicide, does not need to be mechanically incorporated if rainfall or overhead irrigation occurs within 30 days of application. It does leach into sandy soils somewhat but not sufficiently to cause problems with newly planted vines.

When applied at less than 4 pounds of active ingredient per acre, it will often give only partial control of weeds in the thistle and mustard families. It controls a broader spectrum of weed species than trifluralin but is shorter-lived in the soil. As little as 1/8 inch of initial irrigation has given some incorporation, but 1/2 to 1 inch has been better in sandy soil. Although the higher rates of oryzalin are required for high organic matter soils, it will give good control of many weed species when sufficient water is applied.

Oxyfluorfen can be used on young weeds as either a preemergence or postemergence herbicide. It does not leach in the soil and can be applied after planting with no waiting period. The soil should not be cultivated after application or effectiveness is lost. As little as 1/2 inch of irrigation or rainfall for incorporation within two weeks of application will provide optimum weed control in a sandy soil. More may be necessary in heavier soils.

Oxyfluorfen is somewhat volatile when applied to warm, moist soil. If the soil is moist and warm, control will be better if irrigation or rainfall occurs as soon as possible after application.

Oxyfluorfen has a broad margin of safety at the recommended rate. It can be applied from late fall until February 15, at low rates, to control newly germinating weeds or at slightly higher rates for larger weeds. (It cannot be used after December 31 in the Coachella Valley, however.) Rates of 1/8 to 1 pound of active ingredient per acre will control most weed species, depending on weed size. Oxyfluorfen, at rates of 1 to 2 pounds per acre, can be applied in combination with oryzalin for good season-long preemergence weed control in bearing orchards. Broad spectrum annual weed control can also be obtained by combining oxyfluorfen with paraquat.*

Oxyfluorfen is very effective on filaree, *Malva* (cheeseweed), and red maids. Preemergence, it also controls many other annual broadleaf weeds and some grasses.

---

*Restricted material; permit required from County Agricultural Commissioner for possession and use.

Photo by Mary Ann Sall

The management of insects, mites and many disease and weed pests depends largely on the application of pesticides. But pesticides are of little or no value unless applied with proper timing and coverage. Therefore, calibration of equipment is the first step toward proper coverage and pest control.

The first and major part of this section discusses ground-operated sprayers designed to deliver chemicals to the grapevine, along with a description of sprayer setups used to apply chemicals to the ground for weed or soil insect control or subsurface spraying of field bindweed (perennial morning glory).

In the second section a brief report is provided on the limited use of chemical injection systems using sprinkler-irrigation lines. The final section, on pesticide safety, describes the types of pesticides used in the vineyard, the hazards connected with their use and the measures to be taken to protect workers.

Section IX—PESTICIDE APPLICATION AND SAFETY

Contents

On the preceding page:

An over-the-row boom sprayer applies wettable sulfur in a wine grape (Carignane) vineyard in summer near Wasco, Kern County. Purpose of the application is to eradicate existing powdery mildew infections.

There are different ways of applying chemicals properly to grapevines, but regardless of technique the basics are the same. This means that through the proper calibration steps the primary objective—correct nozzle selection for applying the proper amount of material—must be obtained in relation to pressure and gallons-per-minute (gpm) discharge. Also affecting coverage with air-carrier sprayers are the spray pattern introduced into the air stream and the direction of the air produced by the sprayer. Speed of travel through the vineyard is extremely important. Too fast a speed results in only superficial (exterior) coverage; too slow a speed may not only be uneconomical but may cause a leaf shingling effect that blocks spray penetration into the vine's interior.

EQUIPMENT

Ground spraying of vines is accomplished with one of three general types of equipment: (1) the over-the-vine, high-pressure sprayer that covers two complete rows; (2) the air-carrier, single-row sprayer that treats one-half of each adjacent row; and, (3) the air-carrier, double-row sprayer that covers four half rows or two complete rows (Figures 1-4).

Usually the over-the-vine units apply high rates of 150 to 300 gallons of spray per acre, although this sprayer, with nozzle and pressure adjustments, generally can be made to apply as little as 50 gallons per acre (gpa) effectively. Air-carrier sprayers are, for the most part, designed for low volume applications of 10 to 40 gpa. Either rate, high or low volume, can satisfactorily do the job, providing the unit applying the spray is properly engineered, calibrated and operated. Over-the-vine, high-volume, high-pressure sprayers do have one disadvantage: they require frequent refills, and this calls for either having extra manpower and equipment or a high rate of downtime. Also, the unwieldiness of the spray boom can be an operational and transportation problem.

Spray Rate Terminology—Grapes	
Dilute spray	150 to 300 gpa
Semi-concentrate	50 to 100 gpa
Concentrate spray	10 to 40 gpa
Hi-concentrate spray	4 to 10 gpa

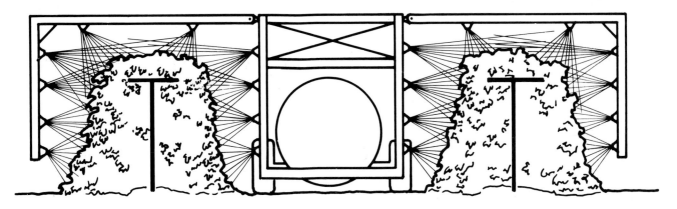

Figure 1. Over-the-vine, high-pressure sprayer.

Some large sprayers are engine-operated; most, however, are powered by the tractor power takeoff (PTO) unit, which usually requires a minimum of 45 horsepower (hp) with 55 to 60 hp preferred. Without sufficient energy input (tractor hp) PTO-operated sprayers do not perform satisfactorily.

While most grape sprayers are mounted on two wheels and towed behind the tractor, some are three-point hitch units or integrally mounted and best suited to the smaller grower or to hi-concentrate applications because of their smaller water carrying capacity.

Tanks

Most tanks are of steel and coated inside with epoxy. Depending on its quality, each coating may last three to five years. If chipping or peeling starts, the inside should be sandblasted and epoxy reapplied before rust or corrosion occurs. Such damage can not only cause screen and nozzle plugging but tank pitting and eventual leakage. Stainless steel tanks have longer wear and are a better investment. Fiberglass also resists corrosion and when properly made and designed, it can be as chemically and structurally sound as stainless steel.

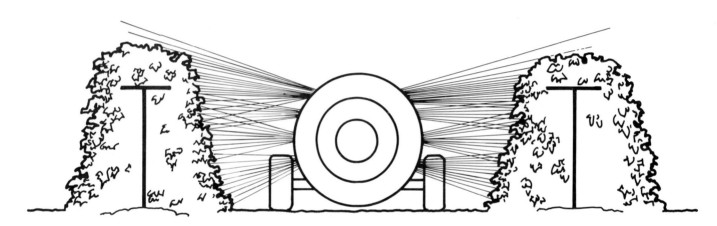

Figure 2. Air-carrier, single-row sprayer.

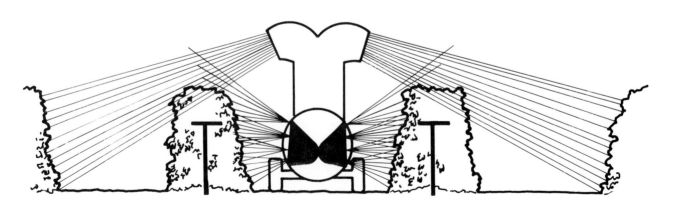

Figure 3. Air-carrier, double-row sprayer covering four half rows.

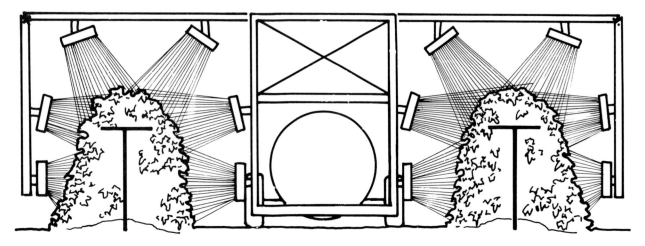

Figure 4. Air-carrier, double-row sprayer covering two complete rows.

Agitation

Continual agitation or mixing of spray material in the tank during sprayer operation is essential. Mechanical or paddle agitators usually do the best job, but some sprayers have hydraulic agitation systems that perform satisfactorily to very good. With this latter system insufficient pump capacity or discharge into the bottom of the tank can produce inadequate turbulence to mix the chemical(s) uniformly, especially when wettable powders are used. This can be even more unsatisfactory when using heavy compounds such as zinc or hard-to-wet materials such as sulfur. It may be necessary to put such chemicals into the tank slowly to get good mixing and suspension or to premix them as a slurry before pouring into the tank.

Pumps

Piston pumps, usually the most expensive, are reliably accurate, positive in action and not too difficult to maintain, once they are understood. They are generally popular. Another type, the roller pump using nylon or rubber rollers, is commonly used where lower pressure and volume are acceptable. Lower cost and ease in parts replacement are favorable factors with this pump.

Nozzles

Gpm discharge is generally governed by the size of the nozzle orifice and/or pressure. Available nozzles vary in type, size and design. Those on high-volume, high-pressure sprayers usually have larger orifices because they must dispense a greater gpm discharge than low volume sprayers. Some spray units use adjustable nozzles, but these quickly lose adjustability and accuracy because of corrosion and wear. Conventional sprayers commonly use nozzle bodies made out of brass or stainless steel and in some designs are fitted with orifice plates made of hardened stainless steel, tungsten carbide or ceramic materials. The latter two resist abrasion and wear best. Nozzle tip wear is greatest with wettable powders.

To provide greater adjustability in gpm discharge, some nozzles have a swirl plate behind the nozzle tip. Its function is twofold: (1) to regulate the gpm that can flow through it and (2) to start the spray in a swirling manner for a hollow or a solid cone pattern.

Spray droplet size is initially determined by nozzle type, orifice size and operating pressure. Nozzle attitude in the air stream and velocity of air also play a part in droplet size in air-carrier sprayers.

Screen filters are necessary on all conventional sprayers. The function of such filters is to screen out debris that may enter the sprayer lines and clog nozzles. Preferably a coarse main screen is located in the line before the pump. A finer mesh screen attached at the base of the nozzle manifold then removes smaller particles. Considerable down time is thereby eliminated by not having to stop the spray operation to clean plugged nozzles.

Another nozzle system, the air shear nozzle, is popular. It is usable only in nonconventional sprayers that operate at very low pressure (14 to 35 psi). Here the nozzle device, made of stainless steel, is placed in the air stream. As the spray liquid is supplied to the nozzle orifice, it is sheared off into droplets by the high velocity air rushing past the nozzle. This application system offers the advantage of little-to-no wear and no plugging. Some air-carrier sprayers do not use nozzles but use instead a simple metering orifice. This has only one function: To release a set amount of spray liquid into the fan discharge area.

Spray Vector

What carries or propels the chemical from the sprayer to and into the vine is the vector. With high-pressure sprayers this is a relatively high volume of water pressurized and emited as a spray from nozzles. In an air-carrier sprayer air is the propelling force utilizing a low volume of water and relatively low pump pressure. The air stream is manufactured by a fan and ducted to carry the chemically-laden water spray to the vine.

Fans in different sprayers vary in size, shape, function and number per sprayer. They also deliver different volumes of air. Air is measured in cubic feet per minute (cfm) as well as in miles per hour (mph) to indicate speed or velocity. Generally, as spray gallonage per acre is lessened the spray droplet's size must also be reduced. To get these lighter, smaller droplets to travel properly to and deposit on the plant, they must in most cases travel at a higher rate of speed (velocity) than larger drops. Therefore, low volume sprayers need to produce high velocity air. The amount of air volume (cfm) produced is not so important since the plant target, the grapevine, is close to the sprayer. This volume, however, may vary from 6,000 to 30,000 cfm.

Sprayer Operation

With over-the-vine, high-volume, high-pressure sprayers, it is important that the gpm discharge be greatest with those nozzles directed at the vine's shoulder. This is the densest part of the vine foliage and where fruit bunches need maximum protection. Better in-the-vine coverage usually results, if the nozzles on one side of the vine point slightly forward toward the line of travel and rearward on the opposite side. This negates the sprays on both sides

of the vine from working against each other and allows easier interior spray penetration. It also is of value on every other nozzle to alternate between a full cone and a more narrow penetrating cone.

Speed of travel in the vineyard is best limited to a maximum of 4.5 mph. The reason for this is the sprayer must remain adjacent to each vine for sufficient time to allow effective coverage on all parts of the vine and especially to the interior area for such pests as omnivorous leafroller, orange tortrix, scale, mealybug and certain disease organisms. Also, excessive speed can damage equipment, particularly on rough ground.

Air-carrier sprayers also need to have the bulk of their air as well as the gpm discharge directed to the vine's shoulder area. This is easily done with air deflectors, nozzle placement and nozzle size. Adjustment may vary for variety, trellising, pruning and other cultural considerations. If possible, the air blast should be directed 15 to 20 degrees forward or 10 degrees rearward from the line of travel. This idea, as with the high pressure, is to cause the spray to enter the vine on an angle to the leaves. This eliminates leaf shingling which otherwise can block spray penetration to the inside of the vine. Speed of travel for single-row, air-carrier units is also best limited to 4.5 mph. Two row and four half-row sprayers should not exceed 3.5 mph because the air-carry distance and penetration considerations require them to remain adjacent to the target for a longer period of time.

Dusters

There are as many different vineyard dusters as there are sprayers. Dusters are mainly used to apply sulfur to control powdery mildew. Coverage requirements for this purpose are generally different from that of insecticides.

Dusts have no special coverage advantages. In fact, they are very difficult to control or direct, are more subject to wind and drift than sprays and for effectiveness are somewhat dependent on plant wetness, leaf hairiness and the amount of horizontal surface area of the target for deposit. It is true that if the pest is susceptible to initial chemical deposit or fuming action, dust can be effective to a greater degree. However, if a residual deposit is necessary for control, only small amounts of dust usually remain on the plant for very long. Therefore, regardless of type of application equipment, the success rate with a dust application is usually low. On the other hand, there are situations that can make such an application necessary and valid.

Unlike certain disease controlling applications, insect dusts need to be applied to every row and at 30 to 50 pounds of dust per acre. Because most dusters have inaccurate flow rates, trial and error methods are needed for metering each different dust formulation. The dust discharge should be directed totally into the vine and not straight up or over the vine tops. Speed of travel is for the most part relative to duster size in regard to discharge and blowing capacity but should not exceed 5 to 6 mph for good penetration and coverage.

CALIBRATING VINE SPRAYERS

The steps in calibrating a vine sprayer are simple and are the same whether the sprayer is a high-volume, high-pressure, low-volume air-carrier, or an herbicide applicator boom sprayer. There are obviously certain things you know about your sprayer operation, and these *knowns* plus a little arithmetic and a nozzle chart will give you the *unknowns* necessary for calibration.

Known:
(1) gpa (gallons per acre) desired
(2) psi (pounds pressure per square inch) desired
(3) mph (miles per hour) desired
(4) number of nozzles on sprayer
(5) vine row spacing

Unknown:
(6) gpm (gallons per minute) needed
(7) nozzle sizes and placement
(8) simple measurement for speed

Following are calibration examples for different vine sprayers:

Example A: Over-the-vine high-pressure boom-sprayer that covers two complete rows per pass (Figure 1).

Known:
(1) gpa = 200 gallons that you wish to apply per acre
(2) psi = 400 pounds pressure at the nozzle manifold
(3) mph = 3-miles-per-hour selected travel speed
(4) number of nozzles = 20 nozzles on sprayer (10 per side)
(5) vine row spacing = 24 feet (Although your row spacing is 12 feet, in spraying two rows,

you must figure for a 24-foot spray swath; for a 10-foot planting, this spray swath would be 20 feet.)

Unknown:
(6) $\text{gpm}^* = \dfrac{\text{gpa} \times \text{mph} \times \text{row spacing}}{500 \text{ (common denominator for formulae)}}$

$$= \frac{200 \times 3 \times 24}{500}$$

$$= \frac{14400}{500}$$

$$= 28.8 \text{ gpm}$$

*Or use the gallons-per-minute chart (Figure 12) to determine the gpm necessary.

(7) There are 20 nozzle positions and you need a discharge of 28.8 gpm.

$$\frac{28.8}{20} = 1.44 \text{ gpm/nozzle}$$

Because vines are not as thick or as dense uniformly from the bottom to the top, it is necessary to use two to three different nozzle sizes (S = small, M = medium and L = large), to place more spray in the shoulder area (Figure 5). To do this, select some nozzles that discharge more and some less than 1.44 gpm. Nozzle calculations are more easily made if based on one row or one side and then later duplicated for the other row. Accordingly, aim for a 10-nozzle arrangement and a total discharge of 14.4 gpm (10 × 1.44 gpm per nozzle). See Figure 6.

A nozzle chart for the type of nozzles on your sprayer (as the following chart shows) will give the gpm discharge of different size tips. Note: if you use swirl plates in your nozzles, be sure to use the nozzle chart corresponding to the swirl plate size.

HOLLOW-CONE NOZZLE CHART

Nozzle Tip No.*	Gallons Per Minute (gpm)					
	50 psi	100 psi	200 psi	300 psi	400 psi	500 psi
4					1.1	
5					1.4	
6					1.9	
7					2.3	
8					2.8	

*Discharge with No. 25 cores (swirl plates).

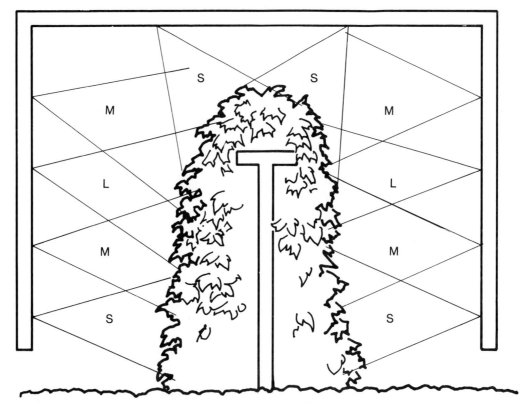

Figure 5. Spray boom on one side of an over-the-vine, high-pressure boom-sprayer showing relative nozzle sizes.

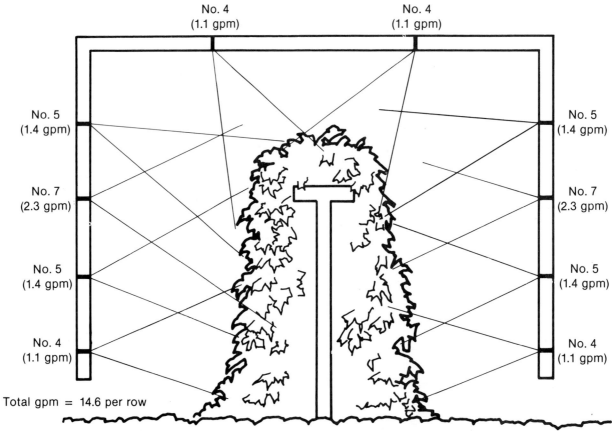

No. 4
(1.1 gpm)

No. 4
(1.1 gpm)

No. 5
(1.4 gpm)

No. 5
(1.4 gpm)

No. 7
(2.3 gpm)

No. 7
(2.3 gpm)

No. 5
(1.4 gpm)

No. 5
(1.4 gpm)

No. 4
(1.1 gpm)

No. 4
(1.1 gpm)

Total gpm = 14.6 per row

Figure 6. Spray boom on one side of an over-the-vine, high-pressure sprayer, showing actual nozzle size selection and placement. This same nozzle arrangement can be duplicated on the other side of the sprayer.

296

For the sprayer in our example, using a spray pressure of 400 psi, check the 400 psi column to find the gpm for each different size nozzle tip. We will need an average output per nozzle of 1.44 gpm, but because of the variations needed for proper vine coverage, we will probably want four nozzles smaller than 1.44 gpm, four nozzles approximately 1.44 gpm and two nozzles larger than 1.44 gpm, placed as shown in the diagram to provide greatest coverage to the densest part of the vine row (Figure 6). Therefore, by selecting four No. 4 tips, four No. 5 tips and two No. 7 tips we can arrive at an acceptable nozzle arrangement.

$$
\begin{array}{l}
4 \text{ No. 4 tips} = 4 \times 1.1 = 4.4 \text{ gpm} \\
4 \text{ No. 5 tips} = 4 \times 1.4 = 5.6 \text{ gpm} \\
2 \text{ No. 7 tips} = 2 \times 2.3 = 4.6 \text{ gpm} \\
\hline
\phantom{4 \text{ No. 7 tips} = 2 \times 2.3 =\ } 14.6 \text{ gpm}
\end{array}
$$

14.6 gpm is satisfactorily close to the 14.4 gpm desired. (Obtaining the exact nozzle gpm output is seldom possible.)

Example B: Air-carrier sprayer traveling each row and spraying two half rows (Figure 2).

Known:
(1) gpa = 50 gallons that you wish to apply per acre
(2) psi = 100 pounds pressure at the nozzle manifold
(3) mph = 3-miles-per-hour selected travel speed
(4) number of nozzles = 10 nozzles on sprayer (5 per side)
(5) vine row spacing = 12-foot row (or swath)

Unknown:
(6) $\text{gpm}^* = \dfrac{\text{gpa} \times \text{mph} \times \text{row spacing}}{500}$

$\phantom{\text{gpm}^*} = \dfrac{50 \times 3 \times 12}{500}$

$\phantom{\text{gpm}^*} = \dfrac{1800}{500}$

$\phantom{\text{gpm}^*} = 3.6 \text{ gpm}$

*Or use the gallons-per-minute chart (Figure 12) to determine the gpm necessary.
(7) There are 10 nozzle positions and you need a discharge of 3.6 gpm.

$\dfrac{3.6}{10} = .36 \text{ gpm/nozzle}$

Because vines are not as thick or as dense uniformly from the bottom to the top, it is necessary to use two to three different nozzle sizes (S = small and L = large) to place more spray into the shoulder area (Figure 7). To do this we must select some nozzles that discharge more and some less than .36 gpm.

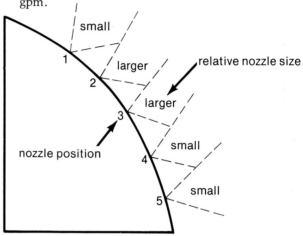

Figure 7. One side of an air carrier showing relative nozzle sizes.

For one side of the sprayer and ease of calculation we aim for a five-nozzle arrangement. Therefore, the five nozzles need to discharge only one-half of the necessary 3.6 gpm, or 1.8 gpm per side.

Using a nozzle chart for the type of nozzles on your sprayer, as the following example shows, check the column headed by the pressure (psi) you are going to use to get the gpm discharge for each different sized nozzle tip. Our example uses a pressure of 100 psi. Note: If you use swirl plates in your nozzle, be sure to use the nozzle chart corresponding to the swirl plate size.

HOLLOW-CONE NOZZLE CHART

Nozzle Tip No.*	Gallons Per Minute (gpm)				
	50 psi	100 psi	150 psi	200 psi	250 psi
2		.25			
3		.29			
4		.45			
5		.54			

*Discharge with No. 25 cores (swirl plates).

Check the 100 psi column for the gpm for each different nozzle tip. In our example we will need an average output per nozzle of .36 gpm, but because of the nozzle size difference for proper vine coverage we will probably want three nozzles smaller than .36 gpm and two nozzles larger than .36 gpm. By selecting three No. 3 tips and two No. 4 tips we can arrive at an acceptable nozzle arrangement sufficiently close to our required 1.8 gpm discharge requirement (Figure 8).

$$3 \text{ No. 3 tips} = 3 \times .29 = \underline{.87 \text{ gpm}}$$
$$2 \text{ No. 4 tips} = 2 \times .45 = \underline{.90 \text{ gpm}}$$
$$1.77 \text{ gpm}$$

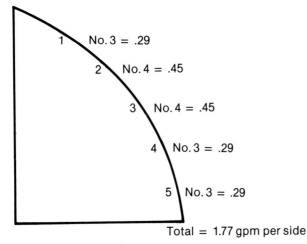

1 No. 3 = .29
2 No. 4 = .45
3 No. 4 = .45
4 No. 3 = .29
5 No. 3 = .29

Total = 1.77 gpm per side

Figure 8. One side of an air carrier sprayer showing actual nozzle size selection and placement. This same nozzle arrangement can be duplicated on the other side of the sprayer.

(8) Those sprayers that use air shear nozzles and calibrate on the basis of gallons per hour would be calibrated as follows: 3.6 gpm × 60 minutes = 216 gph. Since they use low-pressure-air, shear-type nozzles rather than conventional nozzles with different size orifices and cores, nozzle pressure is not important. Therefore, you must follow their simple calibration steps which may consist of merely setting a valve pointer at the desired gph. Still, where adjustment of individual nozzle flow is possible, try to direct the greatest flow to the densest part of the vine as per Figure 9.

Example C: Air-carrier sprayer traveling every other row and covering four half rows per pass (Figure 3).

Known:
(1) gpa = 50 gallons that you wish to apply per acre
(2) psi = 100 pounds pressure at the nozzle manifold
(3) mph = 3-miles-per-hour selected travel speed
(4) number of nozzles = 14 nozzles on sprayer
(5) vine row spacing = 24 feet (Although row spacing is 12 feet, in covering the equivalent of two rows, you must figure for a 24-foot spray swath 2 rows × 12 feet.)

Unknown:
(6) $\text{gpm}^* = \dfrac{\text{gpa} \times \text{mph} \times \text{row spacing}}{500}$

$$= \frac{50 \times 3 \times 24}{500}$$

$$= \frac{3600}{500}$$

$$= 7.2 \text{ gpm}$$

*Or use the gallons-per-minute chart (Figure 12) to determine the gpm necessary for 12-foot row spacing and multiply by × 2.

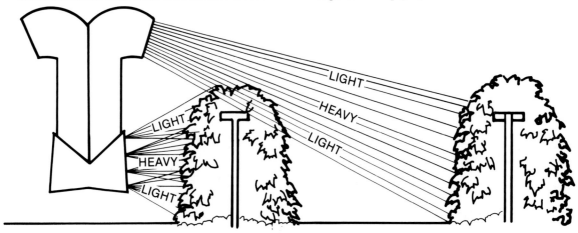

Figure 9. One side of an air carrier sprayer that travels every other row and sprays four half rows.

(7) With this system you are dealing with two air heads per side (each head having its own set of nozzles), one low head for the row adjacent to the sprayer and a high level head directed at the next row beyond the adjacent row (see Figures 3 and 9). In this example you could use three nozzles on the low head and four on the high head. Therefore, you need to be guided by the visual pattern as shown in the preceding figures and by Example B for actual nozzle selection.

(8) Those sprayers that use air shear nozzles and calibrate on the basis of gallons per hour would be calibrated as per the following: 7.2 gpm × 60 minutes = 432 gph. Since they use low-pressure-air, shear-type nozzles rather than conventional nozzles with different size orifices and cores, nozzle pressure is not important. Therefore, you must follow their simple calibration steps which may consist of merely setting a valve pointer at the desired gph. Still, where adjustment of individual nozzle flow is possible, try to direct the greatest flow to the densest area of the vine as per Figure 9.

Example D: Multiple air fan (over-the-vine) two-row sprayers that cover two complete rows per pass and travel every other row (Figure 4).

If the sprayer is an over-the-vine, two-row, air-carrier sprayer utilizing four to six small fans per row, calculate the gpm as per Example A. However, because you are not dealing with high-pressure conventional nozzles you must adjust the spray flow per fan so that the bulk of the spray discharge is directed at the dense shoulder area of the vine. This, as previously mentioned, is where the greatest leaf growth and number of fruit bunches are located. Because these sprayers have adjustable angle fans and may have four to six fans per row, trial and error adjustments are necessary for maximum coverage.

VINEYARD FLOOR SPRAYERS

The function of the vineyard floor sprayer is to cover the entire soil surface between vine rows or to treat vine row berms only (Figures 10 and 11).

Such sprayers can be constructed simply and economically. Small pumps are satisfactory because of the low gallon-per-acre (20 to 100 gpa) and pressure requirements (40 to 60 psi). Straight and

parallel to the ground, booms conduct pesticide and allow proper spacing support for the nozzles which are usually flat fan in design. Because the target (the ground) is relatively flat and at a uniform distance from each nozzle, the same size nozzle can be used across the boom. These nozzles should be so set (distance apart and height above the ground) that they slightly overlap at the edges of their fan pattern. This achieves uniform coverage, as the edges of the fan deposit less spray than the center area. The boom, for safety, should be hinged so that it will compensate if hit when in use and fold back in transporting. With a simple spring attachment it can be so designed that it will return and hold to normal position after retraction.

Calibrating such equipment is simple and can be figured along the same lines as the sprayers directing sprays to aboveground targets. Therefore, if the gpm, psi, mph, number of nozzles and swath width (usually the same as the boom width and/or row width) are known, the unknown can be accurately and quickly determined. For example:

Known:
(1) gpa = 40
(2) psi = 30 or 40
(3) mph = 3
(4) number of nozzles = 7
(5) swath width = 12 feet

Unknown:
(6) gpm* = $\dfrac{\text{gpa} \times \text{mph} \times \text{swath width}}{500}$

$$= \frac{40 \times 3 \times 12}{500}$$

$$= \frac{1440}{500}$$

$$= 2.9 \text{ gpm}$$

*Or use the gallons-per-minute chart (Figure 12) to determine the gpm necessary.

(7) There are seven nozzle positions and you need a discharge of 2.9 gpm.

$$\frac{2.9}{7} = .41 \text{ gpm/nozzle}$$

Because with this type of spraying all nozzles can be the same size, you can look at a flat fan nozzle

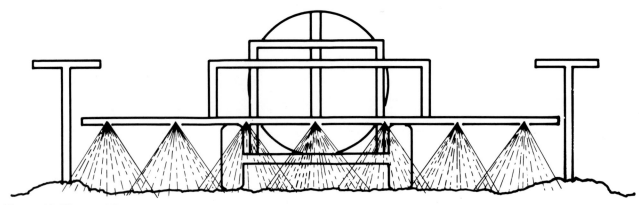

Figure 10. Vineyard floor sprayer with complete coverage boom.

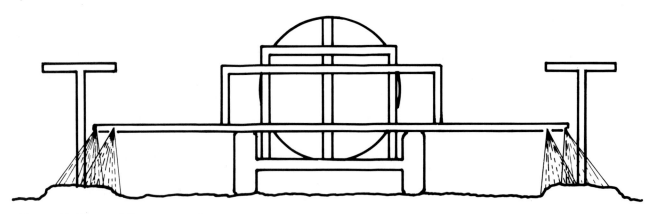

Figure 11. Vineyard floor sprayer for berm only coverage.

chart, as per the following example, and read the output for 30 or 40 psi. From this chart select the nozzle size that most closely discharges .41 gpm. Place seven of these nozzles on the boom (Figure 10), spaced as recommended at 20 inches apart and 17 to 19 inches above the ground.

Note: If spraying just the vine row berms (Figure 11) and you wish to use 2 nozzles per berm, the same procedure is followed. For example:

Known:
(1) gpa = 15
(2) psi = 30 or 40
(3) mph = 3
(4) number of nozzles = 4
(5) swath width = 12 feet (This remains the same even though you are not spraying the whole 12 feet. You are still making a pass through the field (vineyard) every 12 feet.)

Unknown:

$$(6) \quad \text{gpm} = \frac{\text{gpa} \times \text{mph} \times \text{swath width}}{500}$$

$$= \frac{15 \times 3 \times 12}{500}$$

$$= \frac{540}{500}$$

$$= 1.08 \, \text{gpm}$$

FLAT-FAN NOZZLE CHART		
Nozzle Tip No.	psi	gpm
2	30	.17
	40	.20
3	30	.26
	40	.30
4	30	.35
	40	.40
5	30	.43
	40	.50

(7) There are four nozzle positions and you need a total discharge of 1.08 gpm

$$\frac{1.08}{4} = .27 \text{ gpm/nozzle}$$

Because with this type of spraying all four nozzles can be the same size, look at a flat-fan nozzle chart and read down the 40 psi column. From this chart select the nozzle that most closely discharges .27 gpm. Place four of these nozzles on the boom, two at each end for proper berm coverage. Note: Off-center nozzles may be a more desirable selection for the nozzle at the end of the boom in order to eliminate double coverage.

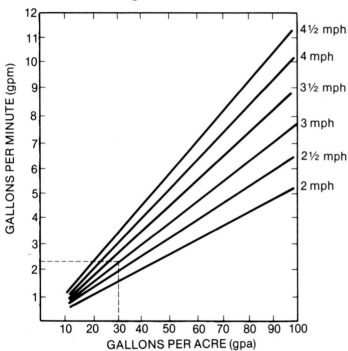

Figure 12. The purpose of this gallons-per-minute chart is to determine the gallons per minute (gpm) necessary at various spray rates per acre (gpa) and speeds of travel (mph) in 12-foot row spacing. This is for air-carrier or boom units spraying one row at a time. Note: Multiply the gpm figure by 2 if using an over-the-vine sprayer that treats two complete rows or four half rows. Also, if spraying more than 100 gpa and up to 200 gpa, plot for one-half the gpa and multiply the gpm amount by 2.

HOW TO USE GALLONS-PER-MINUTE CHART (FIGURE 12)

For example: If the rate you wish to spray is 30 gpa and the speed of travel is 3 mph, the lines drawn on the chart show that a 2.3 gpm nozzle discharge (total for both sides) is necessary. If you have 12 total nozzles (six nozzles per side), then divide 12 into 2.3 gpm and this gives you the gpm discharge per nozzle. By consulting a nozzle chart for your particular type of nozzle and using the column headed by the operating pressure (pressure at nozzle) of your sprayer, you can easily find the gpm of the different size nozzles. Because vines are not as thick or as dense uniformly from the ground up, it is best to use at least two different sizes of nozzles. By doing this you can place more of your spray into the shoulder area of the vines where growth and pests are generally the most dense.

Speed of Travel

You can use the vine spacing to calculate the mph speed of travel on the basis that 1 mph = 88 feet per minute.

Example:

$$\text{Vines/min} = \frac{3 \text{ mph} \times 88 \text{ ft/min}}{8' \text{ vine spacing}}$$

$$= \frac{264}{8}$$

$$= 33 \text{ vines per min} \ (=3 \text{ mph})$$

Or use this table which shows the number of vine spaces you must pass in one minute to obtain your desired (mph) travel speed in relation to the planting distance between vines.

TRAVEL SPEED RELATIVE TO PLANTING DISTANCES

Miles Per Hour (mph)	Number of Vine Spaces Passed Per Minute.		
	Planting Distance		
	6 feet	7 feet	8 feet
2	30	25	22
2½	37	31	28
3	44	38	33
3½	51	44	39
4	59	50	44
4½	66	57	50
5	73	63	55

Spray Check

Finally, to check your calibration:

(1) Spray out an area of vines with plain water to

301

check the spray pattern as well as the total gpa output and rearrange or change the nozzles as needed.

(2) If the gpa is off, check and see whether your original pressure reading (necessary for proper nozzle size selection) was wrong, whether your nozzle size selections were incorrect, or whether speed of travel was too slow or fast.

(3) A better visual method for checking your sprayer coverage is to use a water-soluble dye in your spray tank. Obviously, you must use a dye material that will not injure the vines and that is permissible on grapes. Suitable materials are water-soluble food color dyes used in the baking industry and available in different colors at food supply houses.

For foliage-applied materials, next staple 3×5-inch file cards on a seven-foot lath, one each at one-foot intervals. Place one to three of these laths inside the vines on each side of the travel row against the support wire. Drive the calibrated sprayer past the sampling area and observe the coverage on the cards. Coverage not only shows hits and misses, but also the spray droplet pattern of your sprayer. It is, however, usually best to first make an initial run outside the vine row to see what your pattern looks like in an unobstructed situation. Here you simply anchor the laths 12 feet apart (if 12 feet is your row spacing) and drive the sprayer between the two laths. After reading the results and making minor corrections, then repeat the operation as above in the vine rows. Such a spray check is often revealing when growers use this dye technique with their old existing sprayer setup.

Remember that calibration does not have to be repeated each time a spray needs to be applied. Once properly calibrated for the selected gallonage and speed of travel, coverage will not change. As long as all conditions remain the same and nozzle wear is corrected, the sprayer will function uniformly. Besides nozzle wear, speed of travel should be checked frequently, especially if the operator, tractor or soil surface conditions change.

All spray equipment should be double- or triple-rinsed after use. This involves running clean water through the entire system and nozzles followed with a rich horticultural spray oil/water mix. Leave drain and fill holes open so that tank and lines can dry.

Note: Never treat vineyards with spray equipment in which hormone-type herbicides (2,4-D) have been used; residues can build up in rubber hoses and fittings, leading to subsequent spray contamination and vine damage.

SUBSURFACE SPRAY BLADES

Another special method of application is using a subsurface spray blade for applying the herbicide trifluralin to control field bindweed (perennial morning glory) between the vine rows. This straight blade is passed 4 to 6 inches underground, parallel to the soil surface and injects a uniform layer of herbicide. Nozzles are spaced every 4 to 5 inches apart and spray toward the trailing edge of the blade. The entire assembly is mounted on a tool bar which allows depth control.

Application should be made in 40 to 80 gallons of water per acre. Calibration can be achieved by using either of the methods described under vineyard floor sprayers. Since there is pressure loss in the lines to the blade, pressure should be measured as close to the base of the blade as possible.

GLOSSARY

cfm — cubic feet per minute
gpa — gallons per acre
gph — gallons per hour
gpm — gallons per minute
hp — horsepower
mph — miles per hour
psi — pounds pressure per square inch
PTO — power takeoff

Overvine sprinkler systems have been used for spring frost protection, irrigation and heat suppression, but they have another important use: application of pesticides. Chemicals are injected into sprinkler lines to control certain diseases, insects and mites in vineyards and in some cases water alone may serve to suppress mites.

This technique, while still not widely used, is drawing interest among California growers. Applications, however, are limited to using a few chemicals for specific pests under special local needs registration (see section 24C of the Federal Insecticide, Fungicide, and Rodenticide Act). The only materials presently allowed for use are: wettable sulfur to control powdery mildew, *Uncinula necator;* benomyl (Benlate) and captan (Orthocide) for bunch rot, *Botrytis cinerea;* and carbaryl* (Sevin) for orange tortrix, *Argyrotaenia citrana* (Fernald), and grape leafhopper, *Erythroneura elegantula* Osborn. Use of any other chemicals will require registration clearance and label recommendation.

Successful use of sprinklers for pest control depends on proper delivery and coverage by the system. Methods to meet these needs are outlined in this section. The information presented is from University of California research trials and does not imply recommendations.

Materials and Methods

Adequate coverage of pesticides to grapevines through overvine sprinklers requires a distinct method of delivery involving two separate application timings with a dry-off period between them.

Basically, the required material per acre is put into a pressurized spray tank, such as a conventional speed sprayer, and injected into the main line of the operating sprinkler system. Special procedures must be used because the sprinklers nearest the injection point receive the most chemical for the greatest length of time and sprinklers furthest away receive the least. To equalize these differences, the system is shut off as soon as the chemical in the pressurized spray tank has emptied into the system. This leaves spray solution in the pipelines.

The application is allowed to dry off, usually for one or two hours, so that the material can adhere to the grapevines and minimize wash-off later. Then, the sprinkler system is restarted. Fresh water feeding into the pipelines gradually pushes the remaining chemical solution out through the furthest sprinklers, giving them coverage equal to that obtained earlier by the first sprinklers. But, as soon as the full concentrate is determined to have reached the last sprinkler, the system again is turned off to prevent dilution and wash-off.

Research has shown that the dry-off period in the middle of this split delivery system reduces wash-off to about 20 percent. This is calculated into the per-acre material application rate.

To find out how long it takes for material to disperse through your vineyard system, inject 32 ounces of pure food dye per 100 gallons of water in the pressure-sprayer tank and measure how long it takes to reach the first and last sprinklers. The amount of water and dye used will depend, of course, on the acreage to be covered, the operating pressures and the nozzle opening size of the sprinkler heads.

The injection point, which usually consists of a 3/4-inch injection connector valve on a pressure-sprayer hose hookup, should be located in the main line as near as possible to the acreage to be treated and to the system's pressure pump. It also should be as central as possible to the acreage.

All sprinkler systems will have some differences and require some adaptation, but the following step-by-step procedures are recommended for establishing application times for your vineyard.

(1) Determine the total acreage to be covered per application. When large areas are involved, break them into the smallest units possible for each treatment. Ten to 20 acres is an ideal size, although successful applications on up to 100 acres have been made. Small units can be sequentially treated with the proper adjustments of valves from the main line to laterals.
(2) Fill pressure-sprayer tank with water and connect to the 3/4-inch injection connector valve.

Restricted material; permit required from County Agricultural Commissioner for possession or use.

(3) Start main pump to equalize pressure in sprinkler system.

(4) Start sprayer pump for agitation.

(5) Place the proper amount of dye into the tank and allow time for sufficient agitation to put the material into solution.

(6) Check main line pressure at the injection point and adjust pump pressure at least 20 percent above main line pressure with pump valve closed.

(7) With lateral valves open to acreage being treated and at full psi capacity (normal 55 psi), start the injection by opening the valve at the sprayer pump. Begin timing with a stop watch when opening the valve to the main line.

(8) Record the time it takes for the dye to appear at the first and last sprinkler.

(9) Find out how long it will take to empty the dye solution out of the pressure-sprayer tank. When the sprayer is empty and time is allowed for the dye to disappear at the first sprinkler, shut off the valves in this sequence: first, the main line to the treatment plot; second, the valve leading into the main line from the sprayer; and third, the valve at the sprayer.

(10) Shut off lateral valves in the treatment acreage to prevent backflow; then shut off sprayer.

This ends the first half of the application when determining dye application patterns or, later, for using pesticides. Allow a dry-off period for this first half of the application; the dry-off time should be determined by checking for dry-off in the center foliage of the vines.

After this time interval (one to two hours is normal), turn on the valves to the sprinkler system and main line and, again with a stop watch, time how long it takes to drive the dye from the first sprinkler to the last sprinkler. Shut off the system. That time period is how long the second half of the application should take. This, after a final dry-off period, completes the application.

Special Considerations

Trials using these methods were conducted to study the importance of application timing, wind velocity, grapevine coverage and environmental effects.

Maximum coverage, obviously, was obtained with low wind velocity. Poor coverage resulted when there was significant wind. Low wind velocity of 1 to 2 mph with only a three-minute dye application resulted in more uniform coverage than a nine-minute application with 8 to 10 mph winds. A wind velocity of 4 or 5

mph and a three-minute application showed poor coverage.

Test cards placed on the ground and in vines produced as expected, maximum coverage with 1- to 2-mph winds and nine minutes of application. Cards placed on the ground at a given distance from sprinklers showed a breaking point at an 18-foot distance, diminishing to zero coverage at 32 feet. Spray shift and excessive turbulence occurred with 8- to 10-mph winds, giving poor coverage in some areas. Therefore, proper coverage not only requires a well balanced sprinkler system, but also is influenced by wind and time intervals of application for maximum coverage.

Water applications have varying effects on some mite and insect species. Micro-climate changes produced by sprinkler irrigations may create conditions less favorable for Pacific spider mites to reach their full reproductive potential; moisture conditions greatly alter the reproductive potential of many spider mites. A single sprinkler irrigation will physically remove spider mites from leaves. Mite populations under a sprinkler irrigation program were found to be reduced by approximately one-half. This would indicate possible control of Pacific spider mites by environmental change with overvine applications of water.

Trial results, on the other hand, indicate that grape leafhoppers are favored by overvine sprinkler conditions. There is much lush growth of vines, more host weeds as a food source, and a more readily available water supply. Leafhopper nymphs, however, seem to be less active than adults when subjected to overvine sprinkler irrigation.

Future research may lead to a better understanding of the impact of overvine sprinkler water on insect, mite and disease organisms.

Applying pesticides to vineyards through permanent set sprinkler systems offers important advantages. It almost eliminates pesticide exposure to applicators; in critical periods when immediate coverage is needed, especially for large acreages, total application time is significantly reduced; and field conditions, such as wet soils and vine growth into the avenues that interfere with conventional equipment, are little or no problem.

The major disadvantage, on the other hand, is that *sprinkler applications require both backup safety valves and pumping from a separate water ponding source to avoid any possible contamination of wells*

OVERVINE SPRINKLER APPLICATION
Basic Design

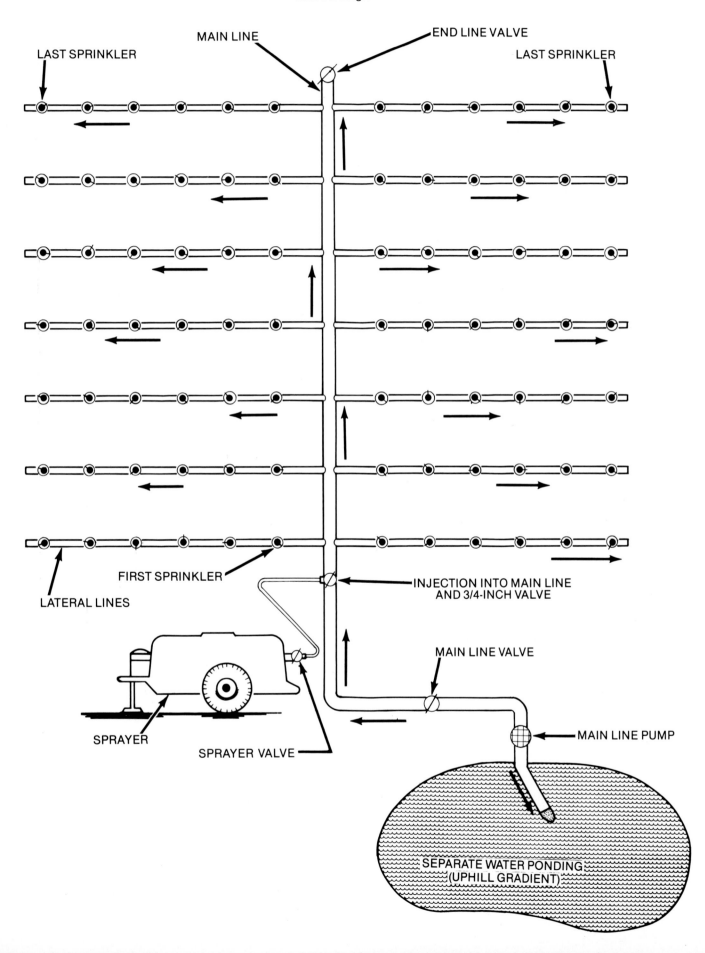

or water sources used for other purposes. This safety requirement must be followed for all pesticide applications. Also, the higher volumes of water used with overvine applications give more runoff, and the final flush-out stage of pipe lines and some wash-off of spray deposits must be considered.

Hillside installations have other problems. Margin rows may have insufficient overlap of sprinklers because of improper pressure delivery in uphill systems not designed to compensate for gradients. Downhill backflow can also cause a major problem during the dry-off period and in final line clearance and application. Special methods are needed to minimize backflow. Valve systems must be adjusted to reduce the area in a given application, more inspection points may be needed, and booster pumps may be required to even out delivery pressures in uphill situations. And, of course, the operator also is faced with the same problems of conventional spray equipment, such as nozzle or sprinkler malfunction, improper water volume and improper timing.

For further information, please contact the authors of this section (see *Table of Contents*).

BASIC OUTLINE OF METHODS USING DYE TO CALIBRATE A SPRINKLER SYSTEM FOR PESTICIDES

Procedure:

(1) Fill sprayer tank with water.
(2) Make hookups from sprayer to main line injection point.
(3) Start main line pump to equalize pressure in sprinkler system.
(4) Start sprayer pump for agitation.
(5) Place the amount of required dye in tank and allow sufficient agitation to put material into solution.
(6) Turn on valves to test area and turn on booster pump.
(7) Check main line pressure at injection point. If the pressure is 60 psi (pounds per square inch) at the main line injection point, set the pressure on the sprayer pump at 100 to 200 psi.
(8) With the valves open in the test area and sprinklers at full psi capacity, start the injection by opening valve to the main line; start timing with a stop watch when main line valve is opened.
(9) Complete injection by allowing the sprayer tank with dye to go empty.
(10) At this point, shut off these three valves in sequence: the main line valve to test area, the valve leading into main line from the sprayer and the valve at the sprayer.
(11) Shut off valves in the sprinkler test area to prevent backflow and shut off the sprayer.
(12) This ends the first one-half pesticide application and leaves one-half of the pesticide in the main line and lateral lines for the final, later application. The above application should give you the number of minutes to the last sprinkler plus two or three minutes hold-on time to the last sprinkler. These additional last few minutes will give coverage to the end vines in the treatment.
(13) Allow a dry-off period of one to two hours. Dry-off should be at the centers of the vines.
(14) After dry-off, turn on valves in the test area. Time your second application from the time you open the main line valve to the test area and turn on the booster pump.
(15) Keep sprinklers at full capacity for _____ minutes*, then shut off the entire system including the valves in the test area. This completes the application.

—Johannes L. Joos
Area Research Farm Advisor

---

*Number of minutes from the first sprinkler to the last sprinkler.

Note: Locations for obtaining dye:
- **Ritz Food Distributors**
  650 Yoland Avenue, Santa Rosa
  Phone (707) 545-4540
  Red food dye 2.5 percent: Need six gallons in 100 gallons of water for sprayer tank.

- **H. Kohnstamm and Co., Inc.**
  2371 East 51st Street
  Los Angeles, California 90058
  Phone (213) 585-1226
  Red dye No. 40: Need one to two pounds in 100 gallons of water for sprayer tank.

The above dye materials are used for color additives in foods.

All pesticides are poisonous to human beings. Therefore, they should be handled cautiously. Improper handling may lead to acute or chronic poisoning and eye or skin damage—even death. Spills can cause fires or explosions.

To learn how to handle a pesticide correctly it is necessary to read the label on its container carefully. The label contains instructions for handling, transporting, storing, mixing and applying the pesticide. Also, the label contains "signal" words that tell you how poisonous the pesticide is. There are three categories of toxicity:

DANGER-POISON (printed in red and usually accompanied by a skull and crossbones symbol) indicates the most toxic compounds. These highly toxic (Category I) pesticides must be handled extremely carefully.

WARNING indicates compounds of medium (Category II) toxicity. Although somewhat less toxic, they, too, must be handled with care.

CAUTION indicates compounds of low (Category III) toxicity. Least poisonous, these compounds must still be handled carefully.

Pesticides are classified according to the pest they control or, in the case of fumigants, according to the way in which they are applied. Insecticides control insects, fungicides combat plant diseases caused by fungi, nematicides are used against nematodes, herbicides are used against weeds, miticides control mites, and rodenticides control rodents. Fumigants are all-purpose, nonspecific pesticides generally used to kill nematodes, fungi and weeds.

Many pesticides can also be classified by one of several chemical groups. Those compounds that are members of the same group have similar characteristics; one important characteristic shared in common within a group is the way in which they can poison man.

Information in the following sections is based on the California Department of Food and Agriculture pesticide safety information series.

INSECTICIDES AND MITICIDES

Organochlorines

Many organochlorines (see chart) can cause acute poisoning if taken by mouth, if spray or dust is inhaled, or if they contact the skin. If any are taken by mouth, nausea and vomiting will occur first; swallowing of a heavy dose may be followed by nervous system malfunction such as twitching, trembling, convulsion and coma. In the case of severe inhalation or absorption by the skin, the nervous system may be affected within an hour.

Although ingestion of relatively small amounts can lead to death, other types of exposure to organochlorines are rarely fatal. However, these chemicals can be stored in fat and with continued daily overexposure, chronic poisoning is possible. Some of these materials are formulated in petroleum solvents which can be more hazardous than the actual organochlorines.

Workers who mix, load or apply pesticides that contain organochlorines should wear the following protective clothing when skin contact with the pesticide is possible:

(1) Clean work clothing (supplied daily by the employer)
(2) Waterproof gloves
(3) Waterproof boots
(4) Waterproof apron if there is a chance of spillage on the body.

Organochlorines Frequently Used in Grapes			
Common Name (Product Name)	Action	Toxicity Category	Re-entry Interval in Excess of 48 Hours
Endosulfan (Thiodan)	Insecticide	II or III[1]	Yes
Dicofol (Kelthane)	Miticide	III	No

[1]Thiodan is usually a Category II pesticide, but in dust formulations containing a low percentage of active ingredient, it is designated as a Category III material.

If there is a possibility of eye injury, goggles or a face shield should be worn.

Organophosphates

Organophosphate chemicals have high pesticidal activity and relatively low persistence in the environment compared with organochlorine pesticides. They are generally highly toxic to humans because of their capacity to affect the nervous system by interfering with the action of the enzyme cholinesterase. Organophosphate poisoning is usually associated with low levels of this enzyme in the blood. Absorption through the skin is a major route of exposure to this type of pesticide, so extreme care must be taken, when handling these pesticides, to avoid skin contact.

Because these materials rapidly enter the body on contact with all skin surfaces and with the eyes, clothing on which there are any significant spills should be removed, and skin should immediately be scrubbed thoroughly with soap and water. Exposed persons must receive prompt medical treatment or they may die. Badly contaminated clothing should be discarded in a safe manner; it cannot be decontaminated.

Symptoms of acute poisoning include: headache, nausea, vomiting, cramps, weakness, blurred vision, pinpoint pupils, tightness in the chest, labored breathing, nervousness, sweating, watering of the eyes, drooling or frothing of the mouth and nose, muscle spasms and coma. There are also indications that long-term low level exposure may damage the nervous system and the eyes. Of the organophosphates mentioned here, very careful attention must be directed toward the use of parathion*, Phosdrin* and Systox*.

Mixers, loaders and applicators working directly with organophosphates should wear the following safety equipment and protective clothing:

(1) Waterproof pants or apron, coat, hat, rubber boots or rubber overshoes
(2) Safety goggles or face shield
(3) Mask or respirator approved by the U.S. Bureau of Mines or National Institute for Occupational Safety and Health (NIOSH)
(4) Heavy-duty, natural rubber gloves.

If any person enters a field that has been treated with an organophosphate pesticide and a re-entry interval exists for that pesticide, the person should be made aware of the re-entry interval. Even if the re-entry interval is in effect, it is legal to enter the field to do certain necessary jobs without protective clothing if contact with plants can be avoided. If there is to be substantial body contact with treated plants, full body protection required for mixers, loaders and applicators must also be provided to the person involved in early entry.

Organophosphates Frequently Used in Grapes				
Common Name (Product Name)	Action	Toxicity Category	Special Dangers	Re-entry Interval in Excess of 48 Hours
Mevinphos* (Phosdrin)	Insecticide	I	Readily absorbed through skin	Yes
Parathion*	Insecticide	I	Readily absorbed through skin	Yes
Demeton* (Systox)	Insecticide	I	Readily absorbed through skin	Yes
Azinphos-methyl* (Guthion)	Insecticide	I	Readily absorbed through skin	Yes
Dimethoate	Insecticide	II	Readily absorbed through skin	Yes
Ethion*	Insecticide	II	Readily absorbed through skin	Yes
Oxydemeton-methyl (Metasystox-R)	Insecticide	II	Readily absorbed through skin	Yes
Naled (Dibrom)	Insecticide	II	Readily absorbed through skin	No

*Restricted material; permit required from County Agricultural Commissioner for possession or use.

*Restricted material; permit required from County Agricultural Commissioner for possession or use.

N-Methyl Carbamates

N-methyl carbamates are members of a larger group called carbamates. However, only those carbamates containing an N-methyl group act on the nervous system by inhibiting cholinesterase. Pesticides containing the more toxic carbamates can cause illness and death if exposure is substantial. These pesticides affect the nervous system in a manner somewhat similar to the organophosphates. Generally, carbamates are fast-acting, but the effects are reversible within six hours. Poisoning symptoms are much the same as with the organophosphates.

Working safely with N-methyl carbamates formulated as powders is difficult. Even with relatively small amounts, the powder can accumulate on the face of the person doing the mixing. A respirator and goggles provide protection for short periods of time, but when sweating occurs, the powder on the face can be carried to the eyes and nose and can lead to poisoning. When using powder or dust formulations, it is very important to protect the eyes, nose and mouth. If exposure does occur, a burning sensation may be felt; more severe symptoms usually develop within an hour. The safest powder formulations are those packaged in water-soluble packets that can be placed directly into the spray tank without opening.

Skin exposure is less of a factor in carbamate poisoning, although it can occur. The major route of poisoning is through inhalation of dust or spray.

Cholinesterase testing must be an integral part of any safety program for those working with carbamates. However, the cholinesterase inhibition caused by N-methyl carbamate pesticides is rapidly reversible, and even though serious poisoning may have occurred, a blood test conducted hours later may show a normal cholinesterase level.

N-Methyl Carbamates Frequently Used in Grapes			
Common Name (Product Name)	Action	Toxicity Category	Re-entry Interval in Excess of 48 Hours
Methomyl* (Lannate, Nudrin)	Insecticide	I	Yes
Carbaryl* (Sevin)	Insecticide	III	No
*Restricted material; permit required from County Agricultural Commissioner for possession or use.			

Both organophosphates and N-methyl carbamates are cholinesterase inhibitors. Under current California pesticide regulations, mixers, loaders and applicators who work with Category I or II organophosphates or N-methyl carbamates for more than 30 hours in any 30-day period must be under medical supervision which includes regular blood testing for cholinesterase levels. When using these pesticides in closed mixing and loading systems for five or more days in any consecutive 30-day period, the employee must be provided with a pre-exposure baseline cholinesterase determination by a licensed physician. These regulations are updated periodically; be sure to keep informed on the latest version from the County Agricultural Commissioner.

When mixing, loading, applying or otherwise handling carbamate pesticides, wear protective clothing, goggles and an approved mask or respirator.

Many pesticides do not fall into the groups mentioned and must be discussed individually. Among these are pesticides not closely related chemically, but which are applied in the same manner. These are the fumigants. In some cases, the term fumigants is a misnomer. A true fumigant is volatile at normal air temperatures; however, some materials classified as fumigants remain liquid.

OTHER PESTICIDES

Fumigants

Chloropicrin* is used as a fumigant both alone and in combination with methyl bromide*. It kills a variety of soilborne organisms, but is also highly toxic to humans and belongs to toxicity Category I. Fortunately, even at extremely low levels, it is easily detectable because of its odor and tear gas effect. Death due to pulmonary edema may result from a lethal exposure. There is a re-entry interval for workers.

The chemical **1,3-dichloropropene***, (Telone II, D-D), is an active ingredient in many fumigants used to control nematodes. It is a Category I material. Skin and eye contact with the liquid can result in severe irritation and absorption if this pesticide is not immediately removed. Inhalation, ingestion and skin absorption can lead to headache, nausea, vomiting, coughing, pulmonary edema and hemorrhaging, chemical pneumonia, liver and kidney injury, anesthesia and death. A quantity of the material insufficient

---

*Restricted material; permit required from County Agricultural Commissioner for possession or use.

to produce an odor may nevertheless be sufficient to cause harm. There is a re-entry interval, and persons entering the field before the interval has lapsed should wear protective equipment.

Ethylene dibromide* (EDB) is another Category I fumigant used to control nematodes in the soil. It is less volatile than most fumigants, and thus less likely than most fumigants to cause acute poisoning.

Immediate symptoms of human poisoning could include headaches, dizziness, nausea, weakness, and in severe cases, liver and kidney damage; central nervous system depression can occur within 48 hours after ingestion. When death occurs, it appears to be due to breathing failure or heart failure complicated by excess fluid in the lungs. Breathing excess amounts causes delayed lung damage and excess fluid in the lungs.

EDB is a severe irritant. Contact of the liquid with skin and other tissues can cause redness, swelling, tissue destruction and general toxicity. In laboratory animal studies, it has proved to be carcinogenic, mutagenic and teratogenic, and it interferes with sperm production in some animals. The drug antabuse and the fungicide thiram greatly increase the carcinogenicity of EDB and, therefore, workers must be warned to avoid exposure to EDB within 30 days of taking antabuse or being exposed to thiram.

As with all highly toxic pesticides, it is important that protective clothing and equipment be worn, and the appropriate precautions be followed. Even with a closed system and shank injection, there will be a potentially dangerous concentration of EDB in the air, and half-face respirators must be worn at all times. There is a re-entry period for EDB.

Methyl bromide* is used to kill many soilborne organisms. One of its important uses is as a nematicide. One of the most dangerous pesticides in common use today, it has caused more work-related deaths in California than any other single pesticide. Part of that danger stems from the fact that methyl bromide is an odorless and colorless gas. Because it is not easily detected, it is easy for a worker to unknowingly inhale the vapors. Whenever possible, it should be used in formulations containing chloropicrin* (a tear gas) as a warning agent. When chloropicrin is used with methyl bromide, it should be emphasized to the worker that eye discomfort is an indication of the presence of a potentially dangerous level of methyl bromide.

Because there is no specific antidote for methyl bromide poisoning, prolonged exposure must be avoided at all times. A full face mask using a standard black canister (not a respirator) should be available for emergency protection but should not be used for more than 30 minutes in an atmosphere of three pounds or more methyl bromide per 1000 cubic feet of air. Methyl bromide rapidly penetrates the skin and a face mask can hide the danger since it eliminates the effect of the warning agent. When high concentrations are present, the gas mask should be used only for life-saving rescue efforts of a few minutes duration because the chemical is being absorbed through the skin throughout the exposure period.

Direct contact of the skin with liquid methyl bromide or high concentrations of its vapors may cause itching and prickling of the skin followed by reddening and later formation of vesicles and blisters which heal slowly. Severe burning of the cornea may result from contact of the liquid with the eyes.

The effects of skin absorption and inhalation of methyl bromide are extremely variable. The effects mentioned here may occur singly or in any combination. Their severity is not necessarily an indication of the seriousness of the poisoning. Early symptoms may include blurred or double vision, nausea, vomiting, dizziness, headache, inflammation of the eyes, slurred and hesitant speech, lack of coordination in muscular movement and elevation of body temperature. Fainting spells, unconsciousness or mild depression may also be present. Breathing may become difficult or stop entirely. Trembling, muscular twitching and convulsions are not unusual.

Sometimes the onset of symptoms may be delayed from four to six hours or more. This is very dangerous because an individual may receive a potentially fatal dose, have only mild symptoms at first and may even show improvement for several hours. Serious symptoms may then develop and the individual may die.

As well as the acute poisoning mentioned, there are chronic effects that can develop from long-term, low-level exposure. Repeated skin exposure to low concentrations can produce acnelike pimples on the face, arms, back and chest. Repeated inhalation or absorption through the skin can result in any and all symptoms described for acute poisoning. Also, fatigue and appetite loss may occur, and disturbance of nerve functioning may be so marked that a person may appear to be intoxicated or may undergo personality changes. One single substantial exposure may pro-

*Restricted material; permit required from County Agricultural Commissioner for possession or use.

duce nerve damage effects such as difficulties with vision and walking. These effects may last for years.

To avoid exposure, methyl bromide* must be applied in a closed delivery system. The hazards involved in using methyl bromide are so great, and worker liability so high, that a grower should seriously consider retaining the services of a licensed pest control operator to apply this material.

Fungicides

Benomyl (Benlate) is a carbamate that does not contain an N-methyl group. It is a fungicide. Considerably less toxic than the N-methyl carbamates, it belongs in Category III.

Captan is a Category III fungicide. It is of relatively low acute toxicity, but tests with laboratory animals have shown it to be teratogenic and carcinogenic for some species. Captan is an irritating material. It will cause eye irritation, and eye contact must therefore be avoided. It often causes skin irritation and, in some individuals, may cause sensitization if the skin is accidentally contaminated. It is harmful if inhaled and those handling it should use an appropriate respirator.

Sodium arsenite*, a Category I material used as a fungicide, is an inorganic arsenic compound and is very poisonous especially when taken orally. Skin and eye contact present less of an acute hazard. Acute poisoning can result in serious illness a few hours after exposure. Symptoms include diarrhea and stomach pain. There may be blood in the urine. Severe poisoning can result in brain damage with noticeable changes in speech and behavior. Studies in other industries have shown that workers who have been exposed to arsenic dust have a higher rate of lung cancer than unexposed workers, so sodium arsenite is considered a potential carcinogen. (Most uses of inorganic arsenic have been cancelled; the last remaining legal use is against phomopsis cane and leaf spot (dead arm) and measles in grapes.) To avoid inhalation, a respirator must be worn when working with this material. Excessive skin exposure to sodium arsenite can produce growths on the skin that may later become cancerous. Workers entering a treated field less than 60 days after treatment should avoid direct body contact with treated plants.

Sulfur, used as a fungicide, is a Category III material, essentially nontoxic to humans. It can, however, cause a skin rash, red tearing eyes, wheezing, cough-

ing and nausea. There is a re-entry interval for sulfur.

Herbicides

Paraquat* is a highly poisonous (Category I) herbicide formulated as a liquid and applied as a spray. Ingestion of even less than one mouthful can be fatal. Skin contamination can cause severe irritation if contaminated clothing is not removed and the skin is not washed immediately. Eye contamination can cause damage to the eye. Inhalation of spray mist may cause irritation and nosebleeds.

Swallowed, an amount as small as one teaspoon of paraquat can result in death from suffocation within three weeks because of the effects of paraquat on lung tissue. There is no specific antidote for paraquat poisoning.

Any mixer, loader or applicator handling paraquat who develops a headache should stop working with the material for the day. Any mixer, loader or applicator who develops a bleeding nose, lips or gums, tightness of the chest, shortness of breath, nausea or vomiting, should be taken to see a physician. It is also necessary to take to a physician any worker who splashes paraquat into the eyes or mouth.

Dinoseb*, a Category I material, belongs to those herbicides known as dinitrophenols. Pesticides containing dinitrophenol are highly toxic to humans and animals. Human poisoning can occur as a result of oral, dermal or respiratory exposure. Since dinoseb is excreted slowly, its toxic effects are cumulative.

Common early symptoms of poisoning are profuse sweating, headache, thirst and weakness. Workers should be alerted to watch for these signs. Warm, flushed skin and fever indicate serious poisoning. The higher the fever, the more serious the intoxication. Nervousness, restlessness, anxiety, unusual behavior or unconsciousness suggest brain injury. Acute poisoning can result in death. Death, which is usually the result of heart failure or respiratory failure, usually occurs within 24 hours after exposure. Weight loss can be observed in persons poisoned with low dosages over a period of weeks.

It is important to understand that this material can be absorbed through the skin very rapidly, resulting in serious kidney, liver and brain damage. Yellow stains on the skin and hair signify that contact has occurred. Yellowing of the whites of the eyes and staining of the urine indicate absorption of potentially toxic amounts.

---

*Restricted material; permit required from County Agricultural Commissioner for possession or use.

Appropriate protective clothing must be worn when working in an area where contact with dinoseb* is possible. To reduce the possibility of contamination, pesticides containing dinoseb must be handled through a closed system when used by employees. If a person becomes ill from exposure to dinoseb, exposure to all phenols must be avoided for at least 30 days.

Miticide

Propargite (Omite) is a miticide in toxicity Category III. This product is very irritating to the eyes and skin and must be handled very carefully. A re-entry interval is not currently in effect, but contact with treated foliage should be avoided for a week after application.

Rodenticides

Strychnine*, another inorganic Category I material, is used as a rodenticide. It is applied as a bait, and the possibility of worker exposure is very limited if safety precautions are used.

Zinc phosphide* is also a Category I rodenticide and like strychnine, it is used in bait formulations. There is no specific antidote for zinc phosphide poisoning.

WHEN TO SEEK MEDICAL ASSISTANCE

Medical assistance should be obtained when:

1. The pesticide has been swallowed.
2. High mist or dust concentrations of the pesticide have been inhaled and symptoms of illness such as nausea or headache are present. In the case of the highly volatile and poisonous fumigants, medical help must be sought anytime accidental inhalation has occurred.
3. Pesticides spray, dust or granules get into the eyes. With the organochlorines, EDB*, 1,3-D, captan, sulfur and some other materials that are either of low toxicity or are not easily absorbed through the eyes, the eyes should be washed with a steady stream of cool, clean water for 15 minutes. If after that time, the eyes do not hurt, it is not necessary to see a physician. With the organophosphates, the N-methyl carbamates, methyl bromide*, chloropicrin*, dinoseb, paraquat*, sodium arsenite*, zinc phosphide* and propargite, the same rinsing procedure should be used, and then the victim must be taken to a physician.
4. Significant skin contamination has occurred. In the case of the organochlorines and other materials where skin absorption is not usually one of the major routes of intoxication, a physician should be seen only if a large area of skin has been contaminated.
5. The worker has become ill or unconscious at work.

The container should be taken, along with the victim, to the doctor, with the label still intact, and with any available literature describing the chemicals involved. If this is not practical, one should copy the name of the product, the active ingredient and the EPA registration number, and give it to the doctor. Inform the physician that he may contact a poison control center, the medical coordinator at the California Department of Food and Agriculture or National Pesticide Telecommunications for information on treatment.

Summary

Currently, there are detailed studies under way on a number of the pesticides mentioned here because of suspected chronic effects. As more information becomes available, tighter use regulations to minimize exposure, or, in a few cases, phasing out of some or all uses may be the result. Dimethoate, benomyl, Telone II* and D-D* are among those that may be affected.

Employees who mix, load or apply pesticides should wear the protective clothing and equipment specified on the label or by current pesticide regulations. Often, but not always, the required protection will include a face shield and appropriate body protection, such as a waterproof apron or a suit. Also (with the exception of only some of the least toxic materials, such as sulfur), a respirator or face mask approved by the U.S. Bureau of Mines or NIOSH should be used.

---

*Restricted material; permit required from County Agricultural Commissioner for possession or use.